10639825

WHAT WOULD NATURE DO?

WHAT WOULD NATURE DO?

A GUIDE FOR OUR UNCERTAIN TIMES

RUTH DEFRIES

Columbia University Press *New York*

Columbia University Press
Publishers Since 1893
New York Chichester, West Sussex
cup.columbia.edu

Library of Congress Cataloging-in-Publication Data
Names: DeFries, Ruth S., author.
Title: What would nature do? : a guide for our uncertain times / Ruth DeFries.
Description: New York : Columbia University Press, [2021] | Includes bibliographical references and index.
Identifiers: LCCN 2020015728 (print) | LCCN 2020015729 (ebook) | ISBN 9780231199421 (hardback) | ISBN 9780231553100 (ebook)
Subjects: LCSH: Resilience (Ecology) | Sustainability. | Uncertainty. | Human ecology.
Classification: LCC QH75 .D3947 2021 (print) | LCC QH75 (ebook) | DDC 333.95/16—dc23
LC record available at https://lccn.loc.gov/2020015728
LC ebook record available at https://lccn.loc.gov/2020015729

∞

Columbia University Press books are printed on permanent and durable acid-free paper.
Printed in the United States of America

Cover design: Milenda Nan Ok Lee
Cover image: Sailesh Patel © Shutterstock

For Yami, Asha, and others yet to be born, with hopes for the world you will inherit.

CONTENTS

PROLOGUE

"THIS is not a drill." "This is not a drill." The headline blasted around the world as I was putting the finishing touches on this manuscript. Suddenly, the frightening reality of our interconnected, unpredictable, and uncertain world became palpable. It could choke your lungs and, through a handshake or cough, infect your neighbor. It could bring down the stock market and crash oil prices.

The culprit was a crown-shaped virus. It might have spread from the blood and guts of a wild animal, perhaps a small, scaly, ant-eating pangolin. A pangolin might have picked up the virus from a bat in a wildlife market in China. The virus passed from the infected animal to a butcher or someone shopping in the market. As one person breathed out the virus and another breathed it in through droplets in the air, it traveled on an exponentially expanding web of sickness and death. It had never before circulated among humans. Within weeks, the virus had piggybacked on travelers and spread across the globe.

Ricocheting repercussions from the virus rocked the global economy. Investors got spooked, and the stock market fell into a downward spiral. In countries around the world, governments imposed shutdowns and people could not leave their homes.

With workers in factories unable to produce goods and ship them around the world, demand for oil tanked. Shoppers hoarded overpriced hand sanitizer and toilet paper. Restaurants and non-essential businesses lost their customers and employees lost their jobs. At this writing, the pandemic has not yet finished its course. The tiny virus has upended the daily lives of nearly everyone on the planet as they struggle to survive through the crisis. Already, hundreds of thousands of lives and trillions of dollars are among the casualties.

The coronavirus confirms that we are all connected in an entangled, complex maze of animals, airplanes, and economies. This is the reality of our modern civilization. Never before has our species been so connected across the continents. Never before have so many of us huddled in cities, dependent on traded goods for our daily sustenance. Epidemics have plagued humanity since civilization began, but today's transport and trade networks give pathogens a conduit to reach nearly everyone on the planet. Humanity has little experience in collectively managing the global reverberations from a microscopic virus in an obscure Chinese market.

Nature's long experience with persisting in an interconnected, uncertain world—the topic I had been pondering for many years for this book—takes on surreal relevance as this book goes to print. If only, as you will read in these pages, humans could cordon off the sick as do bees and termites, the virus wouldn't spread. If only our supply chains resembled a leaf vein's loopy network, hand sanitizer would not be in short supply. If only the stock market could reverse its downward spiral like the self-correcting ebb and flow of insulin in our veins, investors wouldn't be so anxious. Humanity will undergo many more tests as our species encounters new diseases, novel climates, and erratic political whims. We are in uncharted territory. We are not prepared.

By the time this book reaches your hands, probably new, unexpected calamities will have befallen humanity. We have no way to predict exactly when or where the next coronavirus-like shock will strike. We can't say whether it will be another pandemic triggered by a pathogen that jumps from an animal to a person, or whether it will be a calamity like raging fires triggered by the slow-moving train wreck of climate change. We can say with surety that humanity's collective ability to manage our complex world will be on trial. If we put aside our obsessions with efficiency and listen to the wisdom of nature's experience, perhaps humanity can withstand the upheaval and continue to thrive in our complex, capricious world.

WHAT WOULD NATURE DO?

1

THE DRAGONS ARE BACK

JUST below the equator, off the eastern coast of Asia, a long-forgotten mapmaker etched the warning "HC SVNT DRACONES" on a tiny copper globe. Sketches of sea monsters, twisted serpents, and wild beasties adorn the globe, as they do in many medieval maps. The strange creatures signaled uncharted territory to seafaring explorers. Unknown lands, danger, darkness, or perhaps riches lay beyond.

The copper globe, a rare treasure inscribed with the Latin words meaning "here be dragons," resides in the New York Public Library's collection of ancient treasures. It reminds us that there was a time, not so long ago, when uncertainty prowled just beyond the waves. Today, maps chart in fine detail the contours of hills, mountain ranges, the ocean floor, coastlines, and cities. Mapmakers can trace every last reach of the planet with great precision.

But the sophisticated surety of modern maps is deceptive. Dragons still lurk but in a different guise. No one can predict the surprises in store for our hyperconnected, complex civilization, just as medieval explorers could not foresee what lands lay beyond the beasts. Despite vast knowledge about the world, much remains unknown. Precisely when and where a stock market might crash, a political upheaval might upend the norm, a spark might ignite a wildfire, a drought parch the soil, or a deadly disease spread through an unprepared population are impossible to predict. All these disasters lurk in the shadows of the unknown. We live in an age of uncertainty. With a restless planet pushed off-balance from the by-products of an energy-guzzling civilization, and a fragile network that connects nearly everyone on the planet, dragons are all around us.

Modern civilization is ill-equipped and unprepared to navigate through these dragon-filled times. Until the havoc from the coronavirus pandemic hit every aspect of daily life, few paid attention to the warnings from little-known scientists. They cautioned that humanity needs to be on the watch for dangerous viruses that inevitably make the jump from wildlife to human. Other examples of our vulnerability are so common that we hardly recognize them. Engineers design buildings to withstand wind gusts and storm surges experienced within recent memory rather than future possibilities. People purchase homes assuming that forest fires will not threaten their communities, unaware that the future might bring flames to their doorsteps. Ice caps that locked up water for all of human civilization could crash into the sea, flooding coastlines and cities that are home to millions and economic engines for jobs and trade. People in places that become too hot or too dry to eke out food and livelihoods will have little choice but to move their families to more hospitable climes, with little prospect that people in those places will welcome them with open arms.

The human species has no experience living in a world with an atmosphere that has not existed for the last three million years. Even the most sophisticated science and advanced models cannot make up for a basic reality. How our exquisite and ever-changing planet—along with the bounty of life that it harbors—responds to an unprecedented atmosphere is too complex for the human mind to comprehend. Surprises are in store. Unexpected repercussions for diseases, pests, and the plants and animals that provide us with food, water, and clean air are sure to test civilization's capacity to cope.

Another dragon lurks beside the uncertainty about the planet's future from the heat-trapping by-product of energy sucked from long-buried dead plants and animals. People across the world depend on distant places for survival like never before. The story of modern civilization rests on copious energy unleashed from coal mines and oil rigs that replaced the labor of people and animals. With abundant energy, a small handful of people can grow enough food to support many people crowded together in cities. Ships, trucks, and planes can move goods around the world.

Our species has meager experience surviving in our interconnected, urban world. For the vast majority of human existence, people lived scattered around the world, barely aware of one another's existence. Today, billions of people in modern economies rely on a web of dependencies for food, medicines, and other goods essential for survival. A kink in the vast web of trade networks can ricochet and cascade in directions that are not possible to foresee, as we'll see in a later chapter.

The twin dragons of a changing climate and an interconnected economy shape the future. Twentieth-century hubris that our collective knowledge is sufficient to predict and plan is falling by the wayside. Twenty-first-century civilization will persist through nimbleness and resilience founded on the reality of an inherently unpredictable future.

As the recent past no longer serves as a guide for the future, nature's long experience sheds some light. Life in one form or another, whether tiny cells, scaly reptiles, or fur-covered mammals, has flourished for billions of years through catastrophic swings in climate, asteroids crashing into Earth, and mass extinctions. Over deep time and through trial and error, and without intent or preplanned design, surprising tricks to stay nimble and resilient evolved in nature.

Planners, engineers, investors, and governments are unwittingly rediscovering these strategies, from the design of the internet to international negotiations. They are learning that the twentieth century, silver bullet–driven paradigm of efficiency and technological progress cannot protect us from unpredictable fires, storms, market crashes, diseases, and the uncertainty that permeates our complex world.

This book tells stories of these counterintuitive and underappreciated strategies, honed by evolution over billions of years. Collectively they offer a fundamentally different mindset for humanity to persist in the complex, interconnected, and uncertain world of the twenty-first century.

ROME TO TRANTOR

Isaac Asimov, the most prolific science fiction writer of all times, saw it coming. Any complex civilization that becomes too rigid will eventually crumble from encounters with dragons in the form of invasions, political crises, storms, or droughts, among many types of unknowable insults. That's what happened to Asimov's Galactic Empire in his fantastical history of the future some twenty thousand years hence.

Emperor Cleon I presides over the vast galaxy from Trantor, the fictional planet that Asimov introduced to his fans in the first of many novels, *Pebble in the Sky*. Trantor, located at the center of the galaxy, is home to 45 billion people who perform administrative tasks to serve the empire. They live in a giant, dome-covered metropolis that covers the entire planet with highways, living spaces, and human-made structures. Trantorians never see the sky; they are afraid to go outside the dome. Their nutrients come from underground yeast vats and algae farms. The rest of the food comes by spaceship from outer-world planets. The same ships carry away the Trantorians' waste.

Asimov's imagination traveled far into the cosmos, but on Earth he was an imperfect character. He notoriously groped women at every opportunity, grabbed their behinds without their consent, and snapped their bra straps. He liked to work on his copious writings in small, windowless rooms; he rarely left his

New York apartment; and he abhorred travel. He might have felt at home on Trantor. He was also a biochemist, so he knew that a civilization depends most fundamentally on its sources for food and its prospects for disposing of wastes.

The Galactic Empire's central-control strategy does not end well in Asimov's vision. The empire collapses from overreliance on the outer planets and the sheer complexity of ruling the vast galaxy. Asimov took his inspiration from the downfall of the Roman Empire to conjecture a fictional future.

The year AD 476 marked the official end of the Western Roman Empire. Romulus Augustulus, the last of many emperors, surrendered the crown to the Germanic rebel Odoacer. The complex society that had persisted for over a thousand years, with a sophisticated economy, military, and transport system, broke down with no central authority. People fled crumbling cities across the empire, from Roma to Londinium, and retreated to the countryside.

The Roman Empire and other complex societies throughout history functioned through vast networks of connections and dependencies among distant peoples and places. The empire at its height spanned from North Africa to Britain and from Armenia to the Atlantic Ocean. Even common citizens enjoyed material comforts enabled by trade with faraway places. Peasants in central Italy ate from finely crafted pottery manufactured in North Africa and stored their wines and oils in high-quality jars. Graffiti scribbled on buildings and brothel walls hints that many common people could read and write.

Such a sophisticated economy required a high degree of specialized tasks. Some people were skilled potters; others were scribes, professional soldiers, engineers, poets, tax collectors, butchers, bakers, farmers, teachers, and prostitutes, while slaves carried out menial tasks. Some sweated over clay furnaces to make coins from mined silver. The economy was so far-reaching that its ups and downs with wars, plagues, and conquests are to this day recorded in Greenland ice cores. Lead, swept into the air from the furnaces, rained down and froze in layer upon layer of ice.

Trade networks contracted as the complex society dissolved, and material comforts went by the wayside. Tiled roofs and stone floors, common in peasants' houses during the Roman Empire, became a rare luxury. A few decades after the downfall, most people lived in tiny houses with dirt floors and insect-infested thatched roofs. Fine pottery was gone from the kitchens. People no longer scribbled graffiti as literacy became less common.

Historians and archeologists have puzzled for centuries over the causes of the decline of such a sophisticated society. The story sparks fascination because it hits close to home: highly interconnected economy, large cities as hubs, and many specialized tasks; material comforts that far surpassed prior times; strong state-controlled military supported from tax revenues; sophisticated technologies; extensive roads and ports to move goods, people, and armies; and a long-lived civilization that people thought would continue forever. The proximate reason for the downfall rests on Germanic and Hun invaders who infiltrated the empire, sacked Rome, and snatched power from the emperor. But the Roman army had resisted similar attacks two centuries earlier. It could no longer muster the strength. Other reasons made the civilization ripe for decline.

A German historian catalogued from A to Z no less than 210 ways that his colleagues have explained these underlying reasons. Under *C* is Christianity, which burdened society with nonproductive monks and ascetics whose only contributions were abstinence and celibacy, and Corruption as imperial authorities sold favors and military officials extorted money from civilians. Under *P* is Plagues, with waves of bubonic plague that killed so many people that dead bodies clogged city streets. The list goes on: Earthquakes, Excessive Urbanization, Hubris, Soil Exhaustion, Taxation, and Two-Front War. Historian Bryan Ward-Perkins sums up the situation: "Because the ancient economy was in fact a complicated and intertwined system, its very sophistication rendered it fragile and less adaptable to change."

Asimov could have chosen any of history's complex societies for his science fiction view of the distant future. Each has its own mystery shrouded under vines in thick jungles overgrowing ancient carvings, desert sand blown over remnants of abandoned cities, or mountaintops adorned with elaborate temples. Each complex society has its own art, architecture, cuisine, culture, and unique story of its pathway to decline. But they all share common elements, from the sophisticated, urbanized Indus Valley civilization that thrived many millennia ago to the Anasazi who abandoned their iconic cliff dwellings in the American Southwest as recently as the thirteenth century. They all had stratified societies, bureaucracies, and cities supplied with surplus food from the hinterlands. All depended on trade and connections to make society function. Dependencies made them efficient, but those dependencies also made them vulnerable. They were all complex, as measured by the numbers of specialized jobs, trade routes, or reliance on others for the necessities of everyday life.

Political power grabs, enemy invasions, plagues, or a drought or flood could trigger cascading shocks throughout the connected, complex systems. Any one of these triggers could bring down networks that distributed food and essentials to people in the cities or cut off an authority's supply of tax revenue. In the case of the lowland Mayan, for example, the trigger might have been drought that accelerated the long, slow decline set in motion by a rigid, urban ruling elite who resisted change. Steady decline allowed the trigger to spark a collapse.

As in ancient Rome, complex societies brought great benefits, as people could engage in culture-building tasks rather than procure their own food. Each family or small group of people did not have to separately learn how to grow grain, store and mill it, and bake bread. The complex system was efficient. But eventually the complexity made the systems brittle. They became locked into rigid, inflexible institutions and traditions. Bureaucracies and entrenched rulers outgrew the ability to adjust when potential triggers to their downfall inevitably occurred. Or perhaps the rulers became intoxicated with power and wouldn't let go even if it meant their own demise.

Today's world is not similar to Trantor. People still go outside. They see the sky. They eat food grown in Earth's soil. Cities do not engulf the entire planet; rather, they occupy only a small portion of land. Nor is today's world completely analogous to the Roman Empire or other historical complex societies. Today there is no central command from an emperor. Science and technology have extracted copious energy from nonhuman and nonliving sources so that only a few need to engage in producing food in industrialized societies. Knowledge to control the spread of plagues far surpasses the past. But today's world is highly complex and interdependent, more so than anytime in the past. Most people in the world live in cities, perform specialized tasks, and rely on trade networks for nearly every aspect of their material existence. The efficiency is high and the benefits are enormous. But the vulnerability is great. In the words of Ward-Perkins summarizing the learnings from the fall of Rome, "The main lesson . . . is not some specific panacea that can preserve our civilization forever . . . , but a realization of how insecure, and probably transient, our own achievements are—and, from this, a degree of humility."

We might look with humility to life's success over billions of years on our planet for clues to navigate the dragon-ridden seas of the current day.

FROM CLOCKWORK TO COMPLEXITY

The common ancestor to all life emerged in a primordial soup of chemicals. At that moment, predictable chemistry and physics of the early Earth gave way to unpredictable complexity. A primitive cell, the master foreparent of all whales, grass, humans, trees, fish, bacteria, elephants, microbes—all life on Earth—lived sometime in the first billion years of the planet's 4.5-billion-year history. It thrived in the deep sea where underwater volcanoes vent heat and spew a cocktail of chemicals into seawater.

Once life was under way, the planet and its biosphere were not just complex. The interacting oceans, atmosphere, and life became, in the nomenclature of scientists who study such phenomena, a complex adaptive system. When parts of the system are connected and each is able to adjust to its surroundings, as ancient cells can do, feedbacks cascade and reverberate through the system. The system constantly adjusts and changes unpredictably.

Kevin Kelly, the eclectic writer and astute observer of technology, nature, and organizations, posed a riddle to experts in adaptive systems: What color is a chameleon on a mirror? The camouflaging lizards, with their distinctive ability to change color to blend with their surroundings, might not enjoy the experiment. But the speculations from the experts are revealing. A range of unexpected outcomes ensue from the back-and-forth exchange as the chameleon tries to blend with the color that the chameleon itself is reflecting into the mirror.

One expert asserts that the creature would settle at the middle value in its color range. Kelly himself thought that "the poor beast trying to disappear in a universe of itself would endlessly cycle through a number of its disguises." Another opined that the colors would fluctuate chaotically in a random, psychedelic paisley. Yet another idea was that both the chameleon and the mirror would freeze into the chameleon's initial color. The possibilities are endless and would likely be different for each chameleon–mirror pair. The chameleon might turn red, which it tends to do when scared, setting off a scarier and scarier escalation as the mirror reflects back a frightened chameleon. Or, because the chameleon and mirror cannot be truly isolated, the image of the observer might get reflected in the mirror and set off another color change in the chameleon if someone is watching to see what happens. Uncertainty rules in a complex adaptive system.

The riddle is fun to think about, though a bit cruel to the chameleon. On a much grander scale than a chameleon perched on a mirror, exchanges between life, the atmosphere, the ocean, and back to life again is the never-ending story of Earth's extremely complex, adaptive, and self-correcting system throughout its long history.

Bacteria's early switch in survival strategy set off cascading feedbacks that changed the course of all life to come. For a billion years, simple bacteria dominated life with their strategy to extract hydrogen from hydrogen sulfide and combine it with carbon and energy from the sun or from deep-sea vents to produce sugars. Then bacteria expanded their repertoire to extract the necessary hydrogen, not from hydrogen sulfide in swamps or in the sea but from ubiquitous hydrogen bonded with oxygen, in other words, water. That seemingly small adjustment cascaded into blue-green algae that revolutionized life.

Oxygen was a by-product of the water strategy for extracting hydrogen. At the time, life was adjusted to the low-oxygen atmosphere. As oxygen, the blue-green algae's waste product, built up in the atmosphere around two and half billion years ago, the feedback was fatal for some. Bacteria accustomed to low-oxygen conditions retreated to airless, stagnant waters. They were locked into their low-oxygen strategy and couldn't adjust. But the rising oxygen that led to their downfall was a boon for other life.

Green photosynthesizing plants thrived from bacteria's disaster as enough oxygen built up in the atmosphere to shield them from harmful ultraviolet radiation from the sun. Flowerless liverworts, so named for their small, liver-shaped waxy lobes, and hornworts with green, spiky tubes, then mosses, ferns, and cone-bearing plants dominated life on land for millions of years. Then sponges, corals, and jellyfish flourished in the oceans, followed by insects, reptiles, dinosaurs, mammals, and other animals on land.

With animals on the scene, plants had more options beyond wind and water to disperse their seeds. Flowering plants could co-opt bees, birds, and butterflies for procreation with the allure of nectar. Wings and feet of mobile insects and birds could deliver male pollen from inside a flower to the female ovule. The brokered match fertilized the plant's seeds, a task that a stationary plant could not achieve on its own. Plants developed vivid colors and shapes to attract pollinators. Pollen that rubbed off on a hummingbird's head when it dipped its long slender bill into a brightly colored flower for a sip of nectar, for example, traveled along when the hummingbird moved on to the next flower. A similar strategy emerged in fruit-bearing shrubs and trees to entice birds, and later rodents, bats, lizards, and any fruit-loving animal, with the juicy flesh of the fruits. Animals eat the seeds along with the fruit's flesh and scatter seeds where they defecate.

The new strategies brought another level of dependency and complexity to life.

Life had to contend not only with the problems it created for itself but with erupting volcanoes and bombardments from space. About 250 million years ago, ashes and gases from colossal volcanoes blocked out sunlight and obliterated most forms of life including trilobites, corals, and other marine creatures. But life was far from defeated and roared back in different forms. Another possible life-busting would-be disaster occurred about 66 million years ago. A massive collision between a comet and the Earth spewed dust into the atmosphere and again blocked the sun's energy. Many life forms didn't survive the calamity, including the dinosaurs. The dinosaur's disaster made way for mammals to dominate.

One species of those mammals, *Homo sapiens*, became a dominant planetary force only a short time ago in evolutionary terms. For most of our species' 300,000-year history, people foraged for edible fruits and seeds and hunted animals for meat. They existed within the planet's complex system, which provided a diversity of plants and animals, stable climate, and resilience that served life well for billions of years.

About twelve thousand years ago, the survival strategy shifted from hunters and gatherers to settled farmers. Surplus cereals underpinned hierarchical societies, as an elite class could control the food stocks. Settlements of groups of people, trade of foods and ideas, communication, labor-saving technologies, and economies with specialized tasks cascaded into human-created complex societies.

An industrial economy, powered by long-buried plants and animals in the form of coal, begat another leap in the complexity of civilization as recently as a few centuries ago. People left the countryside and moved to cities to work in factories, open shops, and provide services to keep the economy humming. Most people are now urban dwellers in our interconnected, modern world. Like the Trantorians in Isaac Asimov's Galactic Empire, an urban dweller depends fundamentally on connections and dependencies for survival. In the hours that an urban dweller works in an office, factory, or business, or lounges on a couch, someone else is growing the food, pumping the water, generating the energy, bringing the food and water and energy to the city, and taking away the wastes. In turn, urban dwellers work at a myriad of jobs that connect with other people and make society function.

An urban dweller lives a life of invisible connections unprecedented in civilization. Of course, this picture is vastly oversimplified. The complexity of all the interactions needed to keep a city-world fed and housed with sufficient water and energy, and without wastes piling up, boggles the mind. The networked, interconnected system is so complex we can hardly comprehend it, much less control it. Our modern, global civilization and its flows of information and goods resemble more a living, breathing superorganism than a predictable, controllable machine.

A machinelike, clockwork view of the world, promoted by the French father of Western philosophy René Descartes and English physicist and theologian Isaac Newton, guided science and rational thought through the seventeenth-century Enlightenment. Descartes and Newton implanted the notion that the sum is equal to the parts. Knowledge and control are gained from breaking each system—the cosmos, the human body, the economy, or an aircraft—into its individual pieces.

In a simple machine, outcomes are predictable. A gear turns a lever that turns another gear. Every time someone flips a switch, the gears and levers turn the same way with the same response. Even a complicated machine with many parts is predictable and reliable. Push the accelerator of a car and the speed will go up by a predictable amount. Not so when complexity enters into the picture. One part of the system affects another that affects another, while the first part adjusts and sets off another reaction. Uncertainty reigns. No predictable levers can guarantee a reliable and repeatable outcome. It's a chess game, except no one can pretend to know all the rules or the opponent's next move.

Laws of mechanics act like clockwork. Laws of life lead to complexity. When complexity is at play, when the sum is greater—or lesser—than the parts, the clockwork view can lead to hubris. In a living system that is constantly adjusting to its surroundings, a change in one cell, organ, species, or ecosystem can set off a chain reaction that propagates across all the connected parts. Life is always vulnerable to unforeseeable outcomes. The same is true for ancient Rome, the fictional Galactic Empire, and modern civilization. A political uprising, a cutoff in supplies from a distant source, or an unfamiliar dry spell or storm can overwhelm a rigid bureaucracy and cascade through society.

Life on Earth withstands such unforeseeable shocks with a few simple strategies. This book is about some of these strategies and the clues they hold for modern civilization to navigate through uncertain times. You will see that people, businesses, governments, and societies are beginning to retool the way they plan, invest, and make decisions based on the twenty-first-century fundamental reality of an unpredictable future. Their new ways of planning and doing business run counter to the twentieth-century, clockwork view that places efficiency above all else and plans around a predictable future. Unwittingly and sometimes accidentally, they are learning that nature's strategies are the only way to survive in a world full of dragons.

Through trial and error, ideas brewing from the scientists who devote their careers to unraveling the secrets of nature's complexity are starting to prove their relevance in the real world. If civilization is to thrive and prosper, the stories in this book of the initially unwelcome and ultimately successful design of the internet, fortuitous demise of the deadly smallpox virus, rescue of the midwestern wheat crop by a faraway wild grass, and hard-fought cooperation among countries to clean up the atmosphere, among others, are only the beginning of new tools in the tool kit for civilization to thrive and prosper through a dragon-filled future.

In short, and following the order of subsequent chapters, some of nature's strategies that are proving their worth for our human-constructed societies: built-in *self-correcting features*, a stabilizing strategy pervasive in nature and adopted by the stock exchange to catch a free-falling plunge in the market; *diversity*, the hallmark of both financial investors and the natural world, to buffer against an unknown future, keep options open, and safeguard valuable knowledge and ideas from coalescing into a globally homogenous stew of culture, cuisine, and ways of viewing the world; the architecture of ubiquitous *networks*, patterned on tiny veins in a leaf, to keep the flow of goods, food, information, and ideas safe from cascading failure and, conversely, to prevent lethal diseases from spreading; and leaders who enable decisions based on *bottom-up* knowledge of local conditions, the way ants and termites build their fabulous structures, rather than top-down impositions from faraway authorities that inevitably backfire. These are nature's time-tested tactics that maintain life through unknown futures and cycles of renewal. As the clockwork world of the twentieth century recedes into the rearview mirror, these strategies hold the keys to our prosperity and persistence in our dynamic, interconnected, complex world.

2

RECOVERY FROM A CRASH

Install Circuit Breakers

MAY 6, 2010, started out as a shaky day for the U.S. stock market. A debt crisis loomed in Europe and investors were anxious. At 2:32 P.M. Eastern Daylight Time, anxiety turned into panic, triggered by a trader who put more than $4 billion worth of futures contracts on the market. The high volume flooded the market and caused prices to plunge. Automated algorithms, programmed to buy low and sell high, executed transactions on timescales of milliseconds. The contracts became toxic as traders tried to liquidate their shares. Panic reverberated through international markets in what looked like an impending freefall. By 2:45 P.M., thirteen minutes after the crash began, the Dow Jones Industrial Average fell close to a thousand points and the market had lost $1 trillion in value. The precipitous drop became infamous in financial history as the Flash Crash.

The day could have ended very badly were it not for the circuit breaker that broke the fall. After the stock market crash on October 19, 1987, regulators had put in place rules to pause trading and give markets a moment to stabilize when prices plummet too drastically. Twenty-eight seconds past 2:45 P.M. on the day of the Flash Crash, high trade volume and falling prices tripped the circuit breaker. Transactions paused for five seconds.

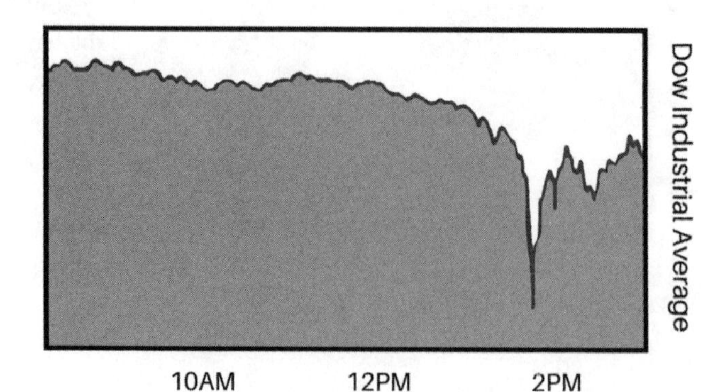

Algorithms recalibrated. The market rebounded almost as fast as it had fallen, and the day ended only 3 percent down from the previous day.

The stock market, like other complex systems of interacting parts, can slip into a downward spiral without a circuit breaker–like, self-correcting mechanism to right itself. With the advent of high-frequency trading and preprogrammed automated algorithms to make buy and sell decisions at lightning speed, the potential for unanticipated outcomes is great. Less intense versions of the Flash Crash continue to occur. Each time, regulators revise the exact drop in value or change in trade volume that defines the circuit breaker–triggering rules.

The coronavirus pandemic put the new circuit-breaker rules to the test on Monday, March 9, 2020. With panic about the economic fallout, and a spat between Saudi Arabia and Russia over who should slash oil production to bring supply in line with demand, panic set in as soon the market opened for the day. By 9:34 A.M., the market had plunged 7 percent, which triggered the rule to halt trading for fifteen minutes. The market continued its downward slide. On Thursday of the same week, confidence was still in short supply. Another 7 percent drop prompted another fifteen-minute break. On Monday, March 16, yet another 7 percent dive tripped the circuit breaker as soon as the market opened for the day, and again on Wednesday, March 18. The circuit breakers might have stemmed a further free fall. In the words of the president of the New York Stock Exchange, the circuit breaker operated "as it's supposed to" to "give investors the ability to understand what's happening . . . and make decisions based on market conditions." As of this writing, the market continues to yo-yo with good and bad news about the possible end to the pandemic's killing spree.

Effective circuit breakers are the golden secret for any complex system—whether a financial market, human body, civilization, or planet—to regulate itself, rein in volatility, and persist over the long term. The potential for unpredictable and erratic behavior in a complex system is too great to leave to chance. On a much longer timescale and with far greater consequences than financial markets, circuit breakers have evolved through geologic time to make our planet conducive to life and life viable on the planet. Human civilization ignores these finely honed circuit breakers at its peril.

A UNIVERSAL SURVIVAL SECRET

When the Earth was very young and covered by oceans, the atmosphere was an oxygen-less stew of heat-trapping carbon dioxide and water vapor that accumulated as volcanoes spewed their gases into the air. Temperatures were scorching hot despite a sun less luminous than today. If gases continued to build up in the atmosphere from volcanic eruptions, liquid water, and with it the chance for life to evolve, would eventually evaporate from the heat. Earth would follow the same fate as our burning hot neighbor Venus with its runaway greenhouse effect. Only a circuit breaker could stop a descent into a boiling hell.

Within a few hundred million years, the churning of material from the hot depths of the Earth to the cooler surface set in motion the single most essential self-regulating circuit breaker that makes all life possible. This churning drives the planetary-scale convection that splits apart ocean ridges where magma rises to form rocks, carries the rocks across the ocean floor over timescales of millions of years, and sends them sliding into the depths at the edges of continental crusts. Collisions where a creeping ocean floor slams into a continent wrinkle the surface to form mountains. The mighty Himalayan Mountains formed when the continental plate carrying what is now the Indian subcontinent

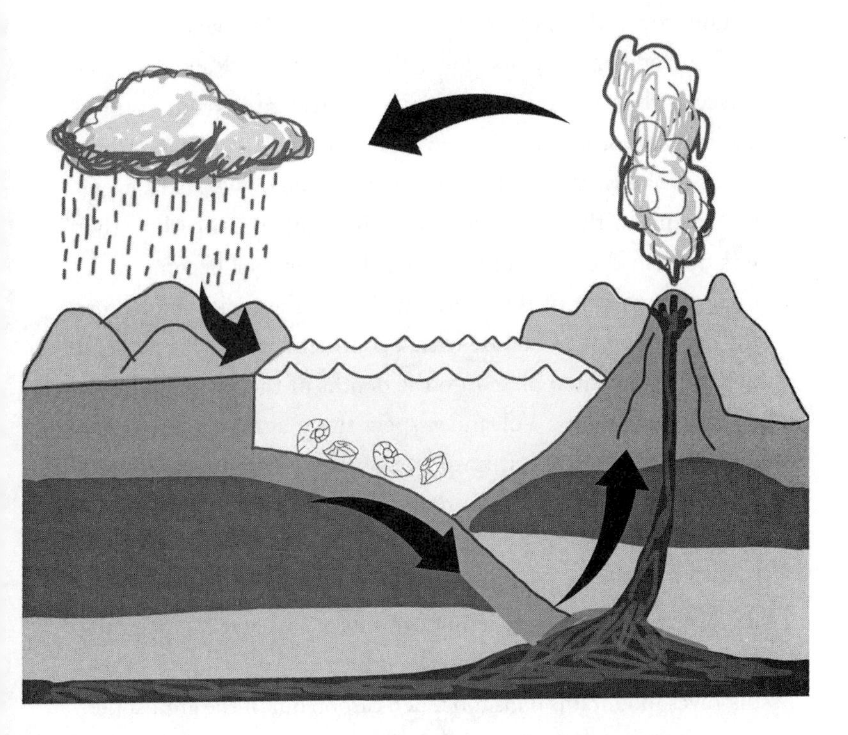

smashed into another plate, and the Rocky Mountains rose when the plate in the Pacific Ocean slid under what is now the earthquake-prone American West Coast.

The marvel of the interacting system of plate tectonics, volcanoes, and the atmosphere lies in its ability to regulate Earth's temperature within bounds conducive to life. The elegant, self-regulating process emerged in Earth's early history and has been operating ever since. The timescales are incomprehensible to the human mind. Mountains erode into mud and continents creep across Earth's surface over many millions of years.

The multistep cycle goes like this: Plate tectonics lift mountains. Carbon from carbon dioxide in the atmosphere dissolves in raindrops, which causes the rain to be slightly acidic. Acidic rain corrodes and wears down the mountains. Rain and dissolved carbon run off the mountains into streams and eventually into the ocean. Tiny organisms in the ocean use the carbon to make shells. The organisms die and some of the shells eventually sink to the ocean floor, carrying with them carbon that was once in the atmosphere. Their dead remains become rocks under the weight of the overlying ocean. The rocks inch across the ocean floor and collide with a continent. The rocks descend under a continental plate into the depths of the Earth and melt to become magma. Volcanoes spew the magma to return the carbon back into the atmosphere. In the meantime, plate tectonics has made more mountains from continental collisions. The cycle continues.

What happens if volcanoes put too much carbon into the atmosphere before weathering can pull it out and temperatures become too hot for life? How could life persist if acidic rain that dissolves mountains pulls too much carbon out of the atmosphere and temperatures are too cold? And, most crucially, how can temperatures remain benign enough for life as the sun becomes hotter over millions of years?

The golden secret is the sensitivity to temperature of the weathering process that grinds down the mountains. Weathering speeds up when high temperatures evaporate more water and more carbon-containing, acidic rain falls from the sky. It slows down when temperatures are cool and would-be rain is locked up in ice. When temperatures are hot, more carbon dissolved in rainfall comes out of the atmosphere. Less carbon dioxide in the atmosphere cools things off. When temperatures are cold, more carbon spewed from volcanoes stays in the atmosphere. As the sun gets hotter, the system self-adjusts with more rain, more heat-trapping carbon pulled out of the atmosphere, and cooler temperatures. Like the trade halt in the Flash Crash, the circuit breaker is in place but with no human-made rules.

Plate tectonics build mountains and weathering wears them down in a cycle that has existed far longer than we can imagine. The temperature control on the pace of weathering sets the bounds and keeps the planet from plunging into lifeless too-hot or too-cold extremes. The elegant, geologic process is no remedy for the carbon going into the atmosphere from the fossil-burning frenzy of modern civilization, unless one is willing to wait millions of years for the weathering process to catch up. But the process has reined in the volatility of wild swings in climate that have plagued other lifeless planets.

Fast-forward through the demise of creatures who thrived in a low-oxygen world and the evolution of plants on land. Abundant oxygen in the atmosphere spearheaded a new life strategy—getting energy to move and grow from eating plants rather than soaking it up from the sun, in other words: animals. Animals could use oxygen inhaled from the air to release usable energy from digested food. The new plant-eating strategy brought mobility for animals, unlike their rooted plant counterparts who needed nutrients from the soil. Most critically, animals could satisfy the copious energy required to maintain a brain. A plethora of animals burst on the scene, sponges, jellyfish, flatworms and roundworms, fish, reptiles, birds, and eventually mammals. Some were cold-blooded and adjusted their body temperatures to their surroundings. Some, namely birds and mammals, were warm-blooded and kept their bodies at a constant temperature that coincides with peak cell performance.

Cold-blooded reptiles, amphibians, and insects are much more efficient than mammals and birds. They get warmth from basking in the sun or lying on a hot rock. At night, when the source of warmth goes away, they simply slow down to conserve energy. Snakes and other cold-blooded animals only need to eat once every few months or even once a year. Humans and other warm-blooded creatures need to generate their own heat to buffer against day-to-night, season-to-season, or place-to-place fluctuations in temperature. We need to eat multiple times a day to generate enough energy to keep our body temperature stable. But the cost of the extra energy for the warm-blooded strategy pays off. We can seek food, defend ourselves, and stay active at night and in a wide range of temperatures. Cold-blooded animals are stuck if temperatures are too low or too high. It's a trade-off, inefficiency of three meals a day versus mobility.

Self-regulating circuit breakers evolved to help resolve the downsides of the warm-blooded strategy. The problem above all others is to keep body temperature at a level constant enough for cells to function. If the temperature is too hot, sensors on the skin send a message to the brain, which in turn sends a message to the sweat glands to produce sweat. Evaporating sweat from the skin brings body temperature down until the sensors send a signal to stop the sweat glands. If the temperature is too cold, the brain sends a signal to the muscles to shiver. The shaking creates heat. As weathering and volcanoes seesaw back and forth to keep the Earth's thermostat at a temperature conducive to life, the brain involuntarily turns on and off sweat glands and shivering muscles to keep our bodies from overheating or suffering a frigid breakdown.

The internal human thermostat is one example of the elegant circuit-breaking, self-regulating strategies that emerged through evolution to offset the shortcomings of warm-bloodedness. Our bodies' sophisticated system that keeps blood sugar within bounds is another. That system counteracts the hazard of the animal strategy of eating food rather than soaking energy from the sun and slurping nutrients from the soil. With variable flows of sugars from food into the bloodstream—a flood during and a drought between meals—the problem is how to maintain blood sugar from climbing too high after mealtime or descending too low between meals. Diabetics know the consequences of an imbalance all too well, confusion, coma, or a trip to the emergency room, and in the longer-term blindness, loss of limbs, and kidney failure.

After a meal when blood sugar is high, the pancreas releases insulin. Insulin in turn ferries sugars to cells and the liver to take them out of the bloodstream. When blood sugar is low, release of insulin slows down, and the pancreas releases another hormone, glucagon, which releases stored blood sugar from the liver into the bloodstream. When blood sugar gets high again, release of glucagon slows down. Insulin kicks in with the next meal and the cycle starts again: back and forth in exquisite homeostasis, oscillating within the bounds of safe levels of blood sugar. The simple, self-regulating system to smooth over potentially dangerous swings in blood sugar from the animal food-eating strategy is not foolproof. Unhealthy high-sugar and high-fat diets can cause havoc with insulin's circuit-breaking role.

The circuit breaker that saved the momentary Flash Crash from spiraling into a stock market collapse on May 6, 2010, the Earth's elegant self-regulating machinery to keep climate swings within bounds, and the seesaw between sweat and shivers and between high and low blood sugar tell the same tale. Built-in mechanisms that keep an inherently unpredictable complex system from spinning out of control are the key to long-term survival. Such stabilizing forces are ubiquitous in nature.

Self-regulating feedbacks inherent in the system—whether a cell, food web, civilization, or planet—hold the golden secret to complex systems that persist over time. The exquisite, life-sustaining feedbacks that evolved through deep time are not easily re-created, as eight brave, oxygen-breathing, food-eating Biospherians found under a glass dome in the Arizona desert.

EARTH BUBBLE

Most everyone alive at the time remembers the moment the first human set foot on the moon. I was at summer camp on July 20, 1969, watching from wooden benches in the mess hall with a roomful of enraptured kids and teenagers. The grainy black-and-white images on television and choppy voice through the speaker captivated the crowd as Neil Armstrong took the famous first step in his clumsy white NASA suit.

Apollo 11, the spacecraft that carried Neil Armstrong and Buzz Aldrin to the moon and back, was an engineering marvel. More than four hundred thousand engineers, scientists, and technicians designed and tested the multiple modules that separated from the spacecraft for a smooth lunar landing. The task was complicated, but it wasn't complex like the interacting system of plate tectonics, volcanoes, and atmosphere that together act as the Earth's thermostat or the hormones that regulate blood sugar. The spacecraft did exactly what the computer code instructed.

No one understood the obedience to code in a nonadaptive system more than the pioneering software engineer Margaret Hamilton. An error in the checklist turned the landing radar switch in the wrong direction, which overloaded the computer with data and set off an alarm just before landing. Hamilton's foresight saved the mission. She had wisely programmed the computer to prioritize the calculations needed for landing rather than the low-priority tasks received from the radar. But neither the spacecraft nor the code learned and adapted as a living being would in such a dangerous situation. The same code run a thousand times would produce the same result unless someone changed the checklist. It was complicated clockwork but not lifelike complexity. Margaret Hamilton's clever piece of code rescued the moonwalk.

It's an amazing feat to get to the moon, stay there for a few hours, and return to Earth. The spacecraft for the eight-day journey was equipped with enough food, oxygen, and other necessities for the astronauts to survive. It's quite another undertaking for a colony of people to live indefinitely in space by growing crops and recycling water and wastes. One person consumes roughly three times his or her body weight in food, four times that weight in oxygen, and eight times that weight in water, and generates 130 pounds of feces and 880 pounds of carbon dioxide in a single year. Clearly, the prospect of carrying enough food to eat, water to drink, and air to breathe for a colony of people is out of the realm of possibility, not to mention the difficulty of carrying away the waste. Even a single glass of water would require massive amounts of energy to lift into space. A colony living indefinitely in space or on another planet would need to grow its own food and cycle its water and wastes. Clockwork switches over to complexity.

Stephen Hawking, the popular cosmologist, firmly believed that humans will colonize space sometime in the future. In an interview before he died in 2018, Hawking cautioned that "the human race shouldn't have all its eggs in one basket, or on one planet" and advised that "it is time to explore other solar systems." Sound advice from the world-renowned expert on black holes and quantum theory. But before colonizing space, the unanswered question is whether a group of people can even recreate the life-enabling, self-regulating systems that we take for granted on Earth.

A handful of farsighted programs and experiments have taken the first steps toward Hawking's vision. The clunky name for such projects is "closed ecosystem life support system" or CELSS. "Closed" means, in the extreme, no inputs or outputs of energy, air, water, or information in a completely sealed system of soil, plants, and animals. In practice, experimental proto–space colony systems are open to energy from the sun. The idea is to replicate on a small scale the vast cycling of nutrients, water, and energy that makes life on Earth possible. It's the same notion as a terrarium that a schoolchild might build in a glass container to marvel as water evaporates from the plants' leaves, condenses on the glass walls, and rains back to the soil.

Space programs in Russia, Japan, Spain, Germany, and the United States have constructed CELSS to grow algae, wheat, lettuce, soybeans, potatoes, and tomatoes in closed capsules no bigger than a house. The most grandiose, expensive, and drama-filled attempt to test whether humans can re-create the life-support systems of Earth stands in Arizona's Sonoran Desert, not far from Tucson, under shiny glass domes.

On September 6, 1991, eight Biospherians in jumpsuits passed through the airlock of Biosphere 2 in Oracle, Arizona. The four men and four women, ranging in age from 29 to 67, had undergone a rigorous selection process. They had herded cattle on a ranch in Australia for six months, swabbed decks, and went scuba-diving from an ocean vessel to demonstrate their technical skills and mental fortitude. They needed both to survive the two-year experiment to live, grow their own food, and run the equipment at Biosphere 2 without stepping into the outside world.

The ambitious idea was bankrolled by Texas billionaire Edward P. Bass who had inherited an oil fortune. Perhaps the plan is better portrayed as outrageous, full of hubris, a publicity stunt, or frivolous. It originated from John Allen, who ran a commune in New Mexico with cultish charisma. The two came up with the plan to construct a self-sustaining mini-Earth, and Bass hired Allen as director.

Preparation and construction took eleven years. Underground, a stainless steel liner isolated tunnels, pipes, and pumps from the surrounding soil. The machinery blew wind, recycled air, and generated waves for the miniature world above. Futuristic glass domes with steel frames sprawled over three acres to house a rain forest, savanna, desert, marsh, ocean with a coral reef, farm, and living and working spaces for the inhabitants. Biospherians collected four thousand species of trees, shrubs, grasses, insects, and edible plants and animals from around the world and implanted them in the artificially constructed biosphere. The idea was that some species would die out, but they would self-organize to sort out which combinations of species could best adapt to their new home. Soil containing countless types of bacteria, fungi, and worms was an essential part of the new biosphere to filter water and recycle wastes. The biosphere was only open to sunlight penetrating through the double-laminated glass and natural gas for energy. The farm was to supply all the food, the marsh to decompose wastes, and the plants in the rain forest and savanna to purify the air for the two-year experiment.

The irony between the technosphere below ground and the biosphere above was not lost on one of the scientists who visited the site. Organic farming and self-sufficiency reminiscent of Allen's commune were as essential to the function of Biosphere 2 as the high-tech machinery. The "earthly smells of compost and forest" were in sharp contrast to the "electronic sterility of the computer control room" and the Biospherians' "simple agrarian lifestyle" to their "sophisticated telecommunications."

The problems began almost as soon as the Biospherians shut the airlock behind them. The eight men and women were hungry despite the copious planning and the many hours they spent farming. Their diet consisted of milk from African pygmy goats; eggs from domestic chickens; meat from goats, chickens, and feral pygmy pigs; tilapia fish from the marsh; and sweet potatoes, bananas, squash, cowpeas, and other fruits and vegetables from eighty plants. But despite the healthy and balanced diet, it wasn't enough to keep their stomachs full and provide their bodies with enough energy to carry out their tasks.

As luck would have it, the first year of the lock-in was cloudier than usual. Less light came through the glass and the crops suffered. Then there was the problem of pests. The Biospherians couldn't use pesticides; the toxins had no place to go in the closed system and would poison the inhabitants. Despite the ladybugs, praying mantises, and wasps brought in to eat pests and careful selection of pest-resistant crops, the Biospherians couldn't keep pests from eating their crops.

The Biospherians lost weight and squabbled over food rations. But, unexpectedly, the shortage of food proved a point. They remained healthy on the restricted but balanced diet. Scientists already knew that an extremely low-calorie but nutrient-dense diet could extend life and prevent age-related diseases in monkeys and rodents. Here was proof that the same was true for humans. The Biospherians were serendipitous guinea pigs.

Breathing was another problem. Oxygen was in short supply as carbon dioxide in the air skyrocketed. A breath was equivalent to huffing at heights more than halfway to the peak of Mt. Everest. And carbon dioxide levels fluctuated widely between night and day. It turned out that the microbes in the rich, organic soil that the Biospherians had spread on the farm were belching carbon dioxide. The plants couldn't take up all that carbon dioxide to produce enough oxygen. At night without sunlight, when the plants were not producing oxygen, the levels of carbon dioxide built up rapidly in the small volume of air in Biosphere 2.

The problem got even worse. Oxygen began to mysteriously disappear from the air. With the health of the Biospherians at stake, management decided to violate the premise of a closed-system experiment. A year and four months into the two-year stint, trucks containing 31,000 pounds of liquid oxygen drove up the access road and injected it into Biosphere 2. The reason for the vanishing oxygen wasn't clear until scientists tested the concrete after the two-year experiment. Carbon dioxide was reacting with calcium in the concrete in the Biosphere's walls, essentially locking up oxygen in the structure.

The desert turned into shrubs with more moisture than expected. Vines overran other plants. Bees and other pollinators went extinct, so the plants could not reproduce. Crazy ants and cockroaches eventually overran the place.

The eight Biospherians emerged, gaunt but triumphant on September 26, 1993, two years after they had entered through the airlock. They exited in two factions of four, neither one speaking to the other. As with other small groups living in close quarters on submarines and spacecraft, they didn't all get along. One group

didn't agree with the decision to violate the bedrock rules of a closed system by pumping in oxygen. The other group was not willing to sacrifice their health to principle. But despite the problems, the opening of the closed system, and the strangling vines, the Biospherians fulfilled the goal to survive two years under a glass dome.

The second mission into Biosphere 2 ran into problems from human foibles rather than concrete walls or ants and cockroaches. A new crew of seven entered through the airlock on March 10, 1994, armed with more pest-resistant crops. On the outside, a storm brewed from mismanagement and cost overruns. The original budget of $30 million ballooned to $200 million. In April, Bass's bankers, accompanied by armed federal marshals and sheriff's deputies, issued a restraining order against management. Bass brought in Steve Bannon, the same Steve Bannon who later brought a wrecking ball to human decency at Breitbart News, to resolve the financial troubles.

In defense of the inhabitants, two of the original Biospherians staged a mutiny from the outside. Abigail Alling, one of the two later charged with a felony for the break-in explains: "On April 1, 1994, at approximately 10AM . . . limousines arrived on the biosphere site . . . with two investment bankers hired by Mr. Bass. . . . They arrived with a temporary restraining order to take over direct control of the project. . . . With them were 6–8 police officers hired by the Bass organization. . . . They immediately changed locks on the offices. . . . All communications systems were changed (telephone and access codes) and [we] were prevented from receiving any data regarding safety, operations, and research of Biosphere 2." The bankers "knew nothing technically or scientifically, and little about the biospherian crew." She judged it her "ethical duty to give the team of seven biospherians [inside Biosphere 2] the choice to continue with the drastically changed human experiment . . ., or to leave . . . It was not clear what they had been told of the new situation." Blocked by the new access codes and unable to talk to the people inside, she and a fellow veteran Biospherian snuck on to the premises, opened the doors of Biosphere 2, and broke the glass seals. "In no way was it sabotage . . . It was my responsibility," Alling said in defense. With lawsuits pending and financial mismanagement, the second experiment ended abruptly and long before schedule on September 6, 1994.

Eventually, Columbia University and then the University of Arizona took over the facility for research. In the words of scientists looking back on the grand experiment, "It proved impossible to create a materially closed system that could support eight human beings with adequate food, water, and air for two years." The unknowns are too great, the surprises too many, and the complexity of Earth's real biosphere too vast to fully comprehend. The prospect of re-creating a livable biosphere in a desert, much less on a distant spaceship, proved unattainable even with massive cost overruns. Human minds are no match for the teeming, perpetually interacting, and continually changing biosphere that has evolved exquisite self-correcting circuit breakers to keep its atmosphere in check and its inhabitants nourished.

Life on Earth took billions of years and many experiments to fine-tune its mechanisms. It's no wonder that human societies, who have lived on the planet for a fraction of an instant compared to all life, have a tendency to overlook the life-enabling, self-correcting feedbacks in favor of short-term expediency.

SMOKEY BEAR'S BLUNDER

Human decisions disrupt and ignore life's self-regulating feedback mechanisms at civilization's peril. The most obvious display in modern times is the unceasing buildup of carbon dioxide in the atmosphere with little regard for the climate-altering consequences. Fossil fuels formed from dead remains of ancient plants and animals, pressed and mashed into coal and oil over millions of years, are marvelously efficient when burned to release their pent-up energy. Organisms buried long ago power our cars, power plants, factories, and nearly every aspect of modern life.

Fossil fuels would be a superb solution to previous civilizations' backbreaking and inefficient ways of getting energy from live plants, animals, and human labor, but this efficient solution has a major problem. There is no circuit breaker to take the by-product, carbon dioxide emitted into the atmosphere, out of the air at the same pace it is going in. The result is civilization-altering heat waves, inundated shorelines, and disruption to a reasonably predictable and benign climate that underlies modern civilization's systems to grow food, live in cities, and stay safe from storms. Perhaps the future brings circuit breakers in the form of sophisticated technologies to take the place of geologic-scale weathering. Or perhaps it is sheer hubris to think that human knowledge can replace such a fine-tuned, essential, self-regulating mechanism.

Closer to the ground, the dangers of disregarding self-regulating mechanisms honed over eons is on display in the pine-covered American West, eucalyptus woodlands of eastern Australia, dense shrubs in Mediterranean Europe, and other fire-prone places around the world. The point hit home on a recent hike with my family in a state park in southwestern Colorado. The snow-capped mountains, sweeping vistas, and fresh air were spectacular.

On the way back to the parking lot at the edge of the park, we walked past a sign that said a new housing development was under construction. Several sprawling, suburban ramblers were already part-way constructed. Here was one more expansion into what some call the wildland–urban interface. The interface creates a hazard in fire-prone places where just enough rain falls to make the forest flammable from a lightning strike or a careless matchstick. Too much rain, as in the rain forests of the mighty Amazon or the islands of Indonesia, and a spark won't catch fire. Too little rain, as in the Sahara or Gobi Deserts, and not enough plants grow to fuel a fire. Southwestern Colorado falls right into the sweet spot for fires. The thick bark of pine trees evolved to resist fire. The trees even depend on fire to melt waxy seals on their pine cones and release their seeds.

In the fire-prone American West, people move to the wildland–urban interface for the big spaces and scenery. Then they expect authorities to put out an inevitable fire that threatens their lives and property. But fire was a mainstay of the landscape long before a developer decided to build houses there. Clash of expectations and reality. And Smokey Bear, the beloved adorable mascot of the United States Forest Service, exacerbated the disregard for self-regulating feedbacks that reduce risks from fire.

"Smokey Bear died in retirement at the National Zoo here today. He was 26 years old," reported the obituary that ran on November 10, 1976. "Smokey was found by rangers in 1950, an orphaned and badly burned black bear, clinging to a tree in a National Forest in New Mexico." The game warden took the bear into his home, nursed him back to health, and shipped him to the National Zoo in Washington, D.C. There Smokey became a popular attraction and spokes-bear for the Forest Service's widely disseminated public service announcement. Sporting a wide-brimmed hat, pointing his paw, and appearing in posters and on television alongside Bambi, cute chipmunks, and other cherished forest creatures, sketches of Smokey notified the public: "REMEMBER: Only YOU can prevent forest fires! Douse your campfires, stamp out burning trash, and don't throw lit matches and cigarettes out your window." The underlying message implied that all forest fires are bad, regardless of their source, and an enemy to be stamped out.

Smokey endeared himself to the public. He had an entire zip code all to himself to receive adoring letters from his fans. Children could mail letters addressed to Smokey at 20252. The government copyrighted the image and received royalties from Smokey-like stuffed bears and key chains.

Smokey Bear's message, or rather the Forest Service's presumptions about how to keep people and property safe from fire, existed long before the poor cub was rescued in New Mexico. In 1935, the Forest Service had established the 10 A.M. policy. Firefighters were to extinguish all fires by 10 A.M. after the day the fire was spotted. The notion that the best way to control fires was to extinguish them before they got out of hand came to North America from Europe. In Europe, fire was an unwelcome anomaly. *Foresta Regis*—royal forests—existed for game and for timber to build ships and construct buildings. Fires burned up a valuable commodity.

In North America before the European invasion, fire was part and parcel of Native American existence. Fire attracted game as young plants regrew. It kept passages open to navigate through the forest. Most importantly, small fires kept bigger fires in check. Low heat and patchy fires over small areas acted as advance circuit breakers to prevent disastrous fires from burning too hot or spreading too far when lightning struck or someone set a fire. The self-regulating mechanism of low-intensity fires was not lost on the Native Americans.

Australian Aborigines, like Native Americans, managed fire for their benefit. Fire helped them find wild roots, chased out kangaroos to hunt, and afterward provided new, luscious growth for game to feed. Frequent low-intensity fires kept the forests and grasslands open in a pattern like a patchwork quilt. The mosaic of burned patches prevented the buildup of fuel that would feed a big, intense fire. So extensively did the Aborigines use fire that the British-born Australian explorer Ernest Giles proclaimed that they were "burning, burning, ever burning." Despite the time-tested success of the patchwork pattern to keep wildfires from getting out of control, European settlers put a stop to the practice for fear that the fires would damage the settlers' crops and burn fences that enclosed their cattle and sheep.

Frederick Erskine Olmsted, a U.S. official who studied in Germany and returned in the early twentieth century to manage western forests in the early days of the Forest Service, had no patience for those who advocated the "savage's example of 'burning up the woods' to a small extent in order that they may not be burnt up to a greater extent bye and bye." He called the self-regulating mechanism "simple destruction" against the government's purpose to "keep its lands producing timber crops indefinitely," concluding that "we must . . . attempt to keep fire out absolutely." Fire suppression and the 10 A.M. approach became official policy. After all, Olmsted was putting into practice the lessons he had learned in Germany. Smokey the Bear repeated those lessons and convinced the public that no good could come of fire.

The policy was so effective that one Forest Service official a century later noted that "the reduction in ecosystem fire in the West has been one of the most profound and significant human interventions of any that has occurred over the last century." The official recounted that success hit back with a counterintuitive downside. Small trees, sticks, and branches made the forest a tinder box of fuel "beyond levels ever before experienced in these ecosystems."

Unbeknownst to Smokey's adoring public, a rift in the Forest Service began to develop during the bear's lifetime. Starker

Leopold, Berkeley professor and oldest son of Aldo Leopold who authored the bible of American conservation *A Sand County Almanac*, followed his father's deeply held ethic that soils, plants, animals, and water are a "community to which we belong" rather than "a commodity belonging to us."

Starker Leopold's 1963 report *Wildlife Management in the National Parks* turned the German idea of fire suppression on its head. Espousing the goal to restore each park to "the condition that prevailed when the area was first visited by the white man," he argued: "When the forty-niners poured over the Sierra Nevada into California, those that kept diaries spoke almost to a man of the wide-spaced columns of mature trees that grew on the lower western slope in gigantic magnificence. Deer and bears were abundant. Today much of the west slope is a dog-hair thicket of young pines, white fir, incense cedar, and mature brush—a direct function of overprotection from natural ground fires." His remedy was "periodic burning" to mimic the self-regulating mechanism of natural fires, precisely the opposite of the 10 A.M. policy to extinguish all fires.

Over the objections of those who still believed that firefighters should extinguish all fires, the report made strong inroads in the National Park Service. The first fire under a new "let burn" policy blazed on California's Kennedy Ridge in 1968. One researcher later reflected how National Park Service officials and staff "had a difficult time" as "their entire career and belief system [had been] based on putting fires out and a bunch of hippie PhDs from Berkeley come along and say 'you got to let it burn.' It [was] hard for them to grapple with that idea." Even harder was the idea to intentionally set fires to mimic the natural process. But prescribed burning became official policy in 1978 when the Forest Service abandoned the 10 A.M. policy in favor of allowing natural and prescribed fires to burn.

A century had passed before the managers relearned what the Native Americans and Aboriginal Australians already knew. No matter how careful people are with campfires and matches, fires are not completely preventable in fire-prone landscapes such as the American West and the woodlands and grasslands of Australia. Sooner or later, lightning will strike. Better to be prepared for the spark by using the mechanism that has protected these forests for eons. With suppression of all fires, trees and brush and species that are less resistant to fire have a chance to accumulate. Small and low-intensity fires act as circuit breakers against more devastating fires. Buildup of fuels literally adds fuel to the fire. Smokey

Bear's cuteness was irresistible, but his message misled the public. People need to take care to keep campfires and matchsticks from setting off an accidental fire, but not all fires are bad.

Carefully controlled, intentionally set fires made good sense from the forest's point of view. But the public living in the wildland–urban interface wouldn't always agree. When massive, record-breaking fires broke out in the summer of 1988 in Yellowstone National Park, a prescribed burn got out of hand, although a carelessly tossed cigarette had ignited the largest fire. The outcry from the public and politicians was too loud to ignore. William Penn Mott, director of the National Park Service at the time, tried to explain the "positive and pragmatic side of the fires" as a "renewal of the park ecosystems." But his arguments could not outweigh the short-term economic disaster from the loss of tourism revenue. Congressional representatives and senators signed a petition to the president demanding that the "let it burn" policy be rescinded. To this day, housing developments continue to expand into the wildland–urban interface and managers' efforts to prescribe fires are constrained by lack of funds and objections from the public.

Today the public is paying the price for not heeding the circuit-breaking mechanism of small fires that limit damage from bigger ones. Costs of fighting fires have skyrocketed in the American West and Australia as decades of fire suppression make way for big, intense fires. Just a sampling of the headlines tells the story—March 19, 2018: "Devastating Australia Bush Fire Destroys Scores of Homes." February 8, 2009: "Australia's Worst-Ever Wildfires Kill 130." December 5, 2017: "Tens of Thousands Evacuate as Southern California Fires Spread." And near where I hiked with my family on a pleasant afternoon, a fire broke out about a year later and a June 11, 2018, headline read: "Colorado Blaze Continues to Grow."

Climbing temperatures and a dry fall season, combined with the buildup of forest fuel and people living in harm's way, make a deadly combination. I flew into San Francisco on November 19, 2018, to visit family. It was week two of the horrendous haze that hit the city. As the flight landed, the soot in the air was so thick you could almost grab it. A tiny brushfire in the Sierra Nevada hills had exploded into a raging fire. Gale-force winds blew the smoke westward. By the time firefighters could contain the blaze seventeen days after the first spark, scores of people had died, close to twenty thousand homes and other structures had burned, walls of fire forced people to flee on a moment's notice, and billions of dollars in damage piled up from the most destructive fire in California's history.

No sooner had I drafted these words then another spate of life-threatening fires again hit the American West in the fire season of 2019. The Kincaid Fire burned through the brush of Sonoma County, known for its fine wineries, destroying homes as ferocious winds spread the flames. Further south, the Maria Fire and the East Fire burned thousands of acres and sent people scrambling to evacuate their homes. A combination of winter rains that built up the brush and strong winds in the dry season were key ingredients. In the words of one expert on the California fires, "When both of those switches are on, then it suddenly becomes very relevant that California is about three degrees Fahrenheit warmer than it would be without global warming."

The problem spilled over into another realm. Houses plunged into darkness; children had no light to do their schoolwork; and hundreds of thousands of residents who lost power for days on end were reminded how much our daily life depends on power lines that carry electricity to our refrigerators, computers, and cell phone chargers. Pacific Gas and Electric, the state's largest utility, shut down its service to prevent sparks from live wires that could ignite more fires. In the meantime, developers continue to build homes in the wildland–urban interface. As long as planners and homeowners ignore nature's self-correcting strategies and stay out of harm's way, Smokey's blunder will plague those living in fire-prone places.

Since I added the story of the 2019 California fires in this chapter, more fires in the Southern Hemisphere's autumn of 2019 meant yet another update at the next round of proofreading. Unprecedented bushfires began to burn through eastern Australia. With traditional patchwork burning relegated to times before colonial authorities punished indigenous peoples for the practice, an abundance of flammable fuel made conditions ripe for fire to swoop through the landscape. The year 2019 was the hottest and driest in eastern Australia's recorded history, turning the forests into a tinderbox. By the time rains doused the fires, an area the size of Portugal had gone up in flames; thousands of people had lost their homes and dozens of people lost their lives; millions of animals died as the fires ravaged koala colonies; and smoke billowed into the air and circumnavigated the planet.

In the words of Australian ecologist David Bowman, who has studied the fires for decades, "the old approaches are broken." The country needs to "support Indigenous communities in cultural burning, and enable Aborigine fire practitioners to undertake fuel management, and train both Indigenous and non-Indigenous fire managers." The advice comes full circle from the lessons Olmsted learned in Germany in the early twentieth century.

A GOOD CRISIS

No wonder the property owners around Yellowstone National Park whose homes were threatened in the massive fires of 1988 didn't take well to the notion that fires limit fires. It seems counterintuitive and, in any event, their interest is to protect their property in the here and now. Smokey Bear's message syncs with the human psyche. To comprehend, much less organize, rules around circuit-breaking, self-regulating mechanisms that operate over time spans longer than the immediate and places beyond the horizon runs counter to human nature.

Most everyone would agree with short-term solutions to immediate threats. A circuit breaker to stop a freefall in the stock market? Absolutely. Oxygen pumped into Biosphere 2 when the prospect of replicating the real biosphere faded? Necessary. But "let it burn" to reduce fire damage in the future? Hard to make the policy stick, as the Yellowstone public and politicians proved. Healthy diets to avoid overwhelming the elegant blood sugar–regulating mechanism to stem a diabetes epidemic? A difficult sell in the face of enticing, inexpensive alternatives on the shelf. Decisions coordinated across national governments to keep a convenient energy source from disrupting the geologically slow, life-sustaining mechanism of plate tectonics and weathered mountains? A very steep uphill climb.

Ancient civilizations suffered the same lack of foresight, although simplified explanations of blind overuse of resources or climatic catastrophes don't suffice to explain their downfalls. One of the most long-lasting, the three thousand-year Mayan reign in the densely populated heartland of the southern Yucatán Peninsula, eventually dwindled around AD 1500. By the time Cortez arrived in 1524, the place was nearly abandoned and overgrown with thick forest. What happened? No one is quite sure. The question has been debated for more than two centuries. But trade between the city-states fell off, once-thriving temples and city-states stood abandoned, a prolonged drought took hold, sophisticated waterworks to irrigate fields lay unused, and people living in dense populations scattered. Ancient civilizations from Aztec and Angkor to Mesopotamia and Mycenae have risen and fallen, each with its own intrigue of insurmountable political mismanagement, costs sunk in infrastructure that they couldn't maintain, invasions, dry spells, or too many people for too little food. Whatever circuit breakers were in place proved unable to overcome the fall.

Analogies between ancient civilizations and the modern world strain with the realities of the present. Modern civilization has, compared with the rest of human history, achieved prosperity and health on a scale unimaginable just a century ago. Places and people in the modern world are vastly more connected through flows of information and goods than ever before. Knowledge and science replace superstitions and belief in the divine to cure illnesses and solve problems. We can look down on the planet from space, we can move information across the globe in an instant, and we can see the tiniest germ through a microscope.

Modern civilization has self-regulating mechanisms that were less the norm in ancient societies. Democracy, where it exists, votes incompetent and ineffective officials out of office. Legal systems push back on false accusations. Public opinion, when it finds its voice, can break a slide into danger if feedbacks to those in power are in place. When the air becomes dirty enough and rivers contaminated with pollution from factories, public outcry can translate into political will to clean up the mess. The infamous 1952 Great Smog in London caused thousands of deaths from the pea soup fog and propelled Londoners to switch from lethal coal to cleaner ways to heat their homes. In Athens, Pittsburgh, Los Angeles, Mexico City, and other cities, people now breathe cleaner air compared with twentieth-century soot-laden skies. Even in Beijing, once the most polluted city in the world, the government shut down polluting factories and power plants in the second decade of the twenty-first century. Other places—Delhi, Karachi, and Riyadh to name a few—have yet to feel the positive effect of self-regulating feedbacks as people watch their children walk to school through thick layers of smog.

"Never let a good crisis go to waste," wisely advised Winston Churchill, the World War II prime minister of the United Kingdom. His wisdom is as relevant for civilization's baby steps toward building circuit-breaking, self-regulating mechanisms as it is for politics. With its interdependencies and complexities, modern civilization has no shortage of crises. New diseases—Zika, dengue, Ebola, and coronavirus—seem to appear out of nowhere. Smoke from fires, like the horrendous California fires in November 2018 and Australian fires in 2019, blankets cities downwind. Each time, if information and governing bodies are up to the task, circuit breakers devised from a previous crisis can undergo a new test. New rules can improve on the previous ones. The crisis of the 1987 Black Monday stock market crash, for example, spurred financiers to create the circuit breaker that prevented the 2010 Flash Crash from becoming a Terrible Thursday. Stock market regulators keep changing the rules as they learn with each mini-crash. The 2020 plunge from fear of the coronavirus's domino effect on the economy put the new rules to the test. Those rules will undoubtedly evolve from that crisis and the inevitable next one. The U.S. Forest Service took a century to backtrack on Smokey the Bear's message, but it eventually revised the 10 A.M. policy. The Biospherians in the second lock-in learned from the first and brought more pest-resistant crops into their sealed home.

The prospect of some crises is so petrifying that it demands a circuit breaker to be in place before it can happen. Such is the case with a highly unlikely but still possible re-entry of the eradicated smallpox virus from an act of terrorism, a mistake in a lab, or emergence of the deadly virus once again in the wild.

The very slim possibility that a smallpox-like virus could set off a human catastrophe of unimaginable proportion rises from one of humanity's greatest achievements. The deadly smallpox virus no longer exists on Earth, except as stock in ultra-secure laboratories, thanks to a heroic and harrowing story told in a later chapter. With smallpox eradicated as of 1980, people no longer get vaccines, which means they have no immunity in case the highly contagious and lethal disease somehow reappears.

Donald Henderson, the architect of the incredible smallpox eradication effort, spent his later years working to guard against devastation from a bioterrorist attack. Despite the minimal likelihood of a smallpox outbreak, he devoted himself to an emergency response plan should the unthinkable become reality. He implored countries to have sufficient stockpiles of smallpox vaccines and to prepare for a coordinated global response. In one of his last writings before his death, he was not satisfied that a circuit breaker was truly in place: "Not more than 7 or 8 countries . . . have sufficient vaccine to vaccinate as many as 20 percent of their population should emergency containment measures be needed."

Information gathering and governing mechanisms to act upon information serve as the cornerstones of self-regulating mechanisms both in nature and in society. The human body's temperature regulator, for example, relies on sensors in the skin to send a message to the brain to either sweat or shiver. No one needs to think much about the sensors. In society, information gathering and institutions to take action, such as the World Health Organization to disseminate vaccines and forest managers who set prescribed fires, do not appear automatically. They take vision, leaders, and investments. The investments might seem inefficient, wasteful, and unnecessary when times are stable. People hardly notice when they are effective. Diseases don't spread and fires don't burn down entire towns, hardly headline material. But the absence of information and action could cause havoc.

Life on Earth evolved as an operating system that rights itself. Modern civilization has had much less experience and time to evolve its own mechanisms. As humanity is learning with each crisis, timely attention to circuit-breaking, self-correcting actions can be the golden secret to overcome the problems that our twentieth-century achievements created.

Nature's first suggestion for modern civilization has two parts. The first, in simplest terms, is to care for the feedbacks in complex systems that have evolved over eons. Life on Earth has persisted through millions of years of evolution that fine-tuned many ways—from counterbalancing forces in planetary processes to hormones in the human body—that maintain stable conditions conducive to life. Human knowledge cannot predict the full consequences when these feedbacks cease to operate. Be careful when you disrupt them, even if the most efficient path is to ignore the consequences. Avoid the hubristic idea that you know enough to re-create the feedbacks with your medicines and technologies.

Short-term benefits of ignoring self-regulating feedbacks are enticing: cheap and available sugary treats and processed food with long shelf lives and high profit margins; policies that protect property from fires and allow houses to expand into fire-prone places; and economies based on fossil-fuel energy sources that disrupt stabilizing feedbacks in the climate are just a few. Try to resist human nature to prioritize a short-term end run around the self-correcting mechanisms that have evolved over eons. Over time, the wisdom embedded in those mechanisms will likely come to the fore and the consequences of circumventing them will be costly.

As your modern civilization controls and manipulates nature with ever-more-sophisticated technologies, keep in mind that your knowledge of the biosphere's life-enabling secrets is limited. Continue your experiments that follow Stephen Hawking's advice to venture into space, but do so with humility. Expect that your limited knowledge, human foibles, or more likely both will thwart your efforts.

The second part of the suggestion is also simple to say, but not so simple to do. In the complex systems that society creates, build in self-regulating circuit breakers before you might need them. In human-constructed systems, from cities to whole civilizations, sooner or later a disruption will test their stability. Nature's experience shows that built-in self-correcting mechanisms that trigger when conditions go awry is the only way to survive. Your uber-connected civilization is even more complex and connected than in the past. You can't know all the feedbacks or repercussions of your decisions.

With your sophisticated technologies to track trends and see the world from space, you might be able to suspect when a threat to your security is looming. The finance industry took nature's example to heart and established sophisticated systems for revising and re-revising circuit breakers with each market crash. Managers of some of the world's most majestic forests learned what nature and indigenous peoples knew for millennia and eventually changed their put-out-all-fires policies. Society can learn and change. Keep your bureaucracies and governing bodies nimble to take action as humanity's knowledge about the complexity of the world grows stronger. When times are stable, put circuit breakers in place to prepare, as the hero of the campaign to end smallpox, Don Henderson, did at the end of his world-changing career. The effort to build circuit breakers into your human-constructed world seems messy, inefficient, wasteful, and perhaps superfluous in stable times. But it will pay off when a crisis hits.

HEDGES FOR BETS

Invest in Diversity

THE eight Biospherians who roamed the world in search of plants and animals for sustenance during their two-year lock-in under a glass dome in Arizona had some hard decisions before they closed the air hatch behind them. Pygmy goats, chickens, sweet potatoes, bananas, and squash were obvious. They needed those to eat. Then they needed grass for the goats and insects for the chickens to eat; bees and moths to ferry pollen from flower to flower so the squash and other fruits and vegetables would yield a crop; beetles and worms to turn over the soil; and trees and shrubs to sop up the carbon dioxide that belched from the soil at night. The Biospherians couldn't be sure they were making the right decisions or what other species they might require. So they went for a species-packing strategy, bringing more than they thought they actually needed, just in case some couldn't find a foothold or were outcompeted by their neighbors. By the end of the two years, it didn't matter. Ants, cockroaches, and vines had taken over. Bees and moths died out. Without pollinators, the flowering plants couldn't reproduce for another crop. Even four thousand carefully chosen species didn't provide enough options to produce food and purify the air for human life to persist inside Biosphere 2.

In another desert halfway around the world, over 140,000 meticulously labeled and sealed packets of options lay in cold storage. The collection resides nineteen miles south of Aleppo, Syria in a white-washed, ordinary-looking building, one of eleven international banks that store jewels more precious to humanity than emeralds and diamonds. In the 1970s, countries banded together to safeguard seeds from the plant species that feed the world, among them barley, cassava, chickpeas, lentils, maize, millet, rice, sweet potatoes, and wheat. The internationally coordinated gene banks around the world hold more than 750,000 samples from humanity's most important crops. Each sample contains a few hundred seeds. One bank with more than 120,000 samples of wheat and maize is outside Mexico City, and another with more than 130,000 samples of rice varieties is in Los Baños, Philippines. The largest is the one near Aleppo, which holds seeds of barley, wheat, and legumes from the world's drylands.

Gene banks are humanity's answer to one of nature's ancient secrets. In a world where an asteroid can strike or the climate can shift, life can only survive by preparing for the unknown. Species packing is one strategy to be ready, as the Biospherians knew. They kept options open for an unknowable future. Life holds its options not only in millions of species of plants and animals, but within each individual of each species through each particular combination of genes. Gene banks keep genetic diversity available for the future, albeit only for the seeds of species that people choose to deposit—a tiny fraction of the actual diversity of species in nature.

Each sample of seeds, dried and stored in an airtight container, has its unique quality. Some samples might grow better with more or less rain, some might be able to withstand a fungal attack, and some might be more nutritious or more palatable. Each individual seed has its own distinct combination of traits. The genetic library of hundreds of thousands of varieties from different species grown in different places at different times is open to the world's army of plant breeders. Through the slow, laborious process of crossbreeding different varieties or the modern, faster approach of editing genes, plant breeders regularly use seeds from genetic libraries to test for more productive, disease-resistant, heat-tolerant, or better-tasting combinations of genes.

With the constant barrage of disease, changing climate, wars, destruction, and unforeseeable calamities, gene banks are the underappreciated, fundamental building blocks to feed modern civilization. Gene banks provide the seeds to replace lost varieties of crops and morph new ones. They are repositories of ancient crops and human ingenuity that knew how to coax food from grasses, shrubs, and trees. The diversity holds the options for humanity's future food supply. As American scholar of plants and civilization Jack Harlan wrote, "These resources stand between us and catastrophic starvation on a scale we cannot imagine . . . the future of the human race rides on these resources."

As Syria's civil war followed the 2011 Arab Spring, Ahmed Amri, the Moroccan director of the international gene bank in Aleppo, saw warning signs. A power outage and a breakdown of the backup generators would quickly destroy the climate-controlled collection that stores genetic memories from millennia of humanity's experiments with the world's oldest and most essential crops. Already, the center's headquarters had moved from Lebanon to Syria when the 1977 Lebanese civil war threatened the collections. Amri devised a rescue plan. Staff drove the back roads of northern Syria to bring duplicates of their collections to safety in Turkey. They worked furiously to ship the rest of their collection to ultimate safety above the Arctic Circle.

The rugged glacial terrain of the Svalbard archipelago, wedged between mainland Norway and the North Pole, is one of the world's northernmost inhabited places. The backup of all backups for humanity's food supply juts out of a mountainside on Spitsbergen, the westernmost of the archipelago's islands. Opened on February 26, 2008, the "doomsday vault" keeps its collections in the frozen ground at a safe temperature and out of reach from civil wars or bomb explosions. A 425-foot-long tunnel chiseled into stone houses the precious collection: more than a half billion seeds from 780,000 crop varieties, the largest collection in the world of seeds from humanity's food crops. The Svalbard Global Seed Vault provides insurance for other gene banks of the world. Duplicates, harvested from offspring of banked seeds, are safe for thousands of years in the permafrost. Only the depositing gene bank can withdraw the seeds at a later date. No one else can remove, distribute, or study them. The seeds, stored in sealed boxes, are in lockdown mode.

Cary Fowler, the influential force behind construction of the Svalbard vault, considers gene banks "an insurance policy for the world." The Svalbard Global Seed Vault, he explains, was "constructed by optimists and pragmatists, by people who wanted to do something to preserve options so that humanity and its crops might be better prepared for change." Fowler, the "world's seed banker," who was born in Tennessee and spent much of his career in Norway, was well aware of nature's strategies that keep future prospects alive.

As the conflict escalated in Syria, the seeds from Aleppo arrived safely at the Svalbard Global Seed Vault. But the dangers drove the center's international staff out of Syria. The center launched a sister bank in Terbol, Lebanon, and opened a new bank in Rabat, Morocco. In 2015, the Aleppo center requested a withdrawal from the Svalbard Global Seed Vault, the first time any gene bank had made such a request. One hundred and twenty boxes of seeds from 38,073 varieties of wheat, barley, lentils, and chickpea, among others crops, traveled by plane from Norway to Lebanon and Morocco. The seeds went to stock the new banks in Terbol and Rabat.

The intense efforts to whisk the seeds away from Aleppo and get them back from Svalbard paid off, at least for wheat farmers in the United States. Wheat farmers fear the Hessian fly, so named for the Hessian mercenaries from Germany who brought the fly in their straw bedding when they fought for the British in the Revolutionary War. The pest can devour a crop and severely eat into profits. With rising temperatures across the Midwest, farmers in 2016 panicked when the larvae began to survive through the winter. Researchers at Kansas State University, where Amri had studied, found a solution in seeds they received from the restocked gene banks. One sturdy wild relative of wheat that grows wild in the hills around Aleppo, when bred with domestic wheat, had the right mix of genes to resist damage from the Hessian fly. A common Syrian grass, and the foresight of those with the vision to store its seeds, proved to be a defense against the onslaught of climate change in the midwestern breadbasket. Somehow the hardy grass had the right traits to counteract saliva from the Hessian fly that would otherwise turn a wheat plant into a slimy slurry for the toothless flies to eat.

A year later, climate change hit the ultimate backup in Svalbard. The winter of 2017 was exceptionally warm in the Arctic. Heavy rain and meltwater from the snow gushed into the entrance of the Svalbard vault. Luckily, the water didn't reach past the entrance. The collections in the tunnel were safe, but the incident pierced the notion of a completely fail-safe lifeline for humanity.

Modern civilization was not the first to maintain genetic options and lifelines. People have shared and stored seeds since nomadic hunters and gatherers became settled farmers many millennia ago. Even today, in villages around the world where farmers depend on their own seeds rather than seed companies, neighbors share sacks of rice and grains of wheat and set the seed aside for another planting season. In the seventeenth century, the imperial powers of Europe constructed botanical gardens to conserve their plant conquests and commercially exploit the richness of exotic herbs, trees, and flowers from distant lands.

The Russian plant-collecting explorer Nikolai Vavilov pioneered the notion of collections in his quest "to bring the best plant resources of the world to Soviet agriculture." Vavilov's early twentieth-century expeditions took him to Persia, Afghanistan,

Central and South America, the Mediterranean, and Ethiopia. Through meticulous observations of fine distinctions in shape, color, and characteristics of different plant species, he came upon a revolutionary idea. The cereals, fruits, and vegetables that feed humanity arose from ancient human experiments with breeding plants in different places at different times. He called the places with wild plants similar to modern domesticated crops the "centers of origin." Vavilov eventually identified seven such centers. People domesticated potatoes, tomatoes, and strawberries in the Andean Center; durum wheat, cabbage, and lettuce in the Mediterranean Center; barley, millet, and flax in the Abyssinian Center; corn, lima beans, and sweet potatoes in the Central American Center; lentils, alfalfa, and peas in the Southwest Asiatic Center; and soy, peaches, and onions in the East Asiatic Center.

Plants related to domestic counterparts, today known as crop wild relatives, are the backbone of plant breeding to bring new traits into existing crop species. Wild relatives might not seem appetizing. At first blush their genes seem to have little value, but they could be crucial to improve the crop. After all, the relatives of our domesticated crops, like the common Syrian grass that saved the wheat from the Hessian fly, have survived eons of pests and climate variability in the wild. Another example, the wild relative of tomatoes, whose small, hard fruit is not very edible, recently donated fungus- and worm-resistant genes to the cultivated tomato. After World War I, Vavilov used his collection of crop wild relatives to upgrade Soviet wheat and other staples to frost-hardy, drought-tolerant, and disease-resistant varieties. Soviet authorities awarded him the first Lenin Prize for these achievements.

Vavilov later fell out of favor with the authorities. Inheritance of traits through genes, even in plants, conflicted with anti-bourgeois Soviet ideals. Vavilov was sentenced to prison and

Domestic (left) and wild (right) tomatoes

died there in 1943. The seed collection that he had acquired from so many continents lived on at an institute in Leningrad. In the nine hundred-day Siege of Leningrad during World War 2, when German forces surrounded the city and no food or people could go in or out, the institute's staff squirreled the seeds away and took turns standing guard. Several died of starvation, choosing to protect the valuable collection over eating seeds of rice, wheat, and oats that could have been their salvation.

Today, about one thousand gene banks holding more than six million samples are scattered throughout different countries, including the international centers and gene banks maintained by national governments. The largest are the now-named Vavilov Institute in renamed Saint Petersburg, Russia, and the national seed storage laboratory in Fort Collins, Colorado. The facilities that hold the world's insurance policy struggle for support. In many places collections are deteriorating from lack of adequate climate-controlled facilities.

Gene banks house the genetic memory of humanity's edible species and their wild relatives. That leaves millions of plants in the forests, grasslands, and fields that survive in the wild. Their genetic secrets hold promise for a better crop or a healing medicine, particularly at a time when shifting seasons and rainfall regimes present, as Fowler says, "The biggest challenge agriculture has faced since the Neolithic days when it began."

Crop species alone, even with their wild relatives, cannot feed humanity. The Biospherians were well aware that a host of other assisting species—pollinators, nutrient cyclers, and pest eaters—play key roles in producing the food on their dinner plates. Crops and their relatives are only a tiny slice of life on Earth. No vault is big enough or fail-safe enough to house all the parts of the complex system that feeds civilization.

Diversity of plants and animals and knowledge about how to obtain food from them is the safety valve for the inevitable times when parts of the system fail. Diversity keeps options alive when times change. It keeps resilience intact. Throughout the history of life on Earth, diversity climbed upward in fits and starts and built the scaffold for its very persistence through uncertain times.

A CLIMB TOWARD COMPLEXITY

No one knows how many species exist in the world today. The number could be as few as two million, or perhaps as many as eight million. Some estimate as many as one trillion if you include uncounted, unseen microbial life in the oceans and soil. Scientists have identified, catalogued, and given Latin names, such as *Lycopersicon peruvianum* for the wild tomato that provided fungus-resistant genes, to one and a half million species. This number is surely a subset of the true sum. The overwhelming majority of known species on Earth today are insects. Ants, beetles, and butterflies far outweigh the pittance of all species that have backbones and the even smaller number of mammals. In other words, our planet houses an enormous number and variety of species. Only a fraction is on display in a walk through a forest or a snorkel along a coral reef. We are woefully ignorant about what and where they are, much less how they survive and what they do to keep the planet habitable.

One of the great marvels of our planet is how the incredible diversity of life evolved. Darwin's elegant insight that new species emerge as parents pass on genes that fit the surroundings explains the broad strokes. When surroundings become colder, warmer, drier, wetter, or a new food source appears, those genes that help offspring survive pass on to the next generation. Over time, new species evolve and unfit species disappear. The shifting tapestry of species is as old as life itself. About 540 million years ago, with the explosion in possibilities for life after the brain-enabling Great Oxygenation began more than a billion years earlier, the baton passed from one set of dominating species to another. First, a burst of diversity of fish in the ocean, then reptiles and dinosaurs on land, insects, and finally mammals took their turns as the rulers of life, each for millions of years. Along the way lichens, ferns, and then flowering plants reigned in the world of plants and fungi.

Each new species that joined the tree of life brought the potential to eat or be eaten by another species or to develop a mutually beneficial arrangement. Lichens came early in the Cambrian explosion of life and their green masses spread over the surface of rocks. Their hair-like branching root structures permeated into fissures in rocks and dispersed rock-dissolving acids. Acid and water seeped into the fissures and broke apart the rocks, speeding up the weathering process that pulls carbon from the air. The lag shortened between the time volcanoes spewed carbon dioxide into the atmosphere and the time ocean organisms took up the once-atmospheric carbon in their shells. Swings in climate became less severe, paving the way for other forms of life to evolve with a more stable climate.

Tiny fungi played another outsized role in the surge in life's diversity. Early in the evolution of land plants, plants were short in stature, more like moss and less like the shrubs and towering trees that evolved later. Early plants did not have roots to suck nutrients from deep in the soil. They developed a mutually beneficial arrangement with their microscopic fungi counterparts to overcome each of their shortcomings. Fungi could extract nutrients from rocks with their hairy tentacles, but they couldn't produce sugars. Plants produced sugars obtained from the sun's energy combined with carbon from the air in the process of photosynthesis, but they didn't have nitrogen and other needed nutrients. Fungi colonized the plants' cells to extract sugar. They left nutrients for the plants in exchange. Soybeans, peanuts, and other leguminous plants retained this symbiotic arrangement with fungi. Today, fungi form thick sheaths and live in nodules on these plants' thickly matted roots that branch deep into the soil.

Once animals came on the scene, more possibilities opened up. Marine animals and land insects evolved hard shells to protect against predators. Predators responded by evolving stronger teeth. Gazelles evolved longer legs to escape cheetahs. Cheetahs countered by evolving to run even faster. Newts produced a fatal toxin to kill off would-be predators. Newt-eating garter snakes evolved to avoid harm from the toxin. Bacterial parasites began to live in a water flea host and the host evolved to be immune to the invasion. Once flowering plants evolved, a riot of colors and shapes competed to attract insect and animal partners to ferry the plants' pollen. Even trickery and cheating is not beyond

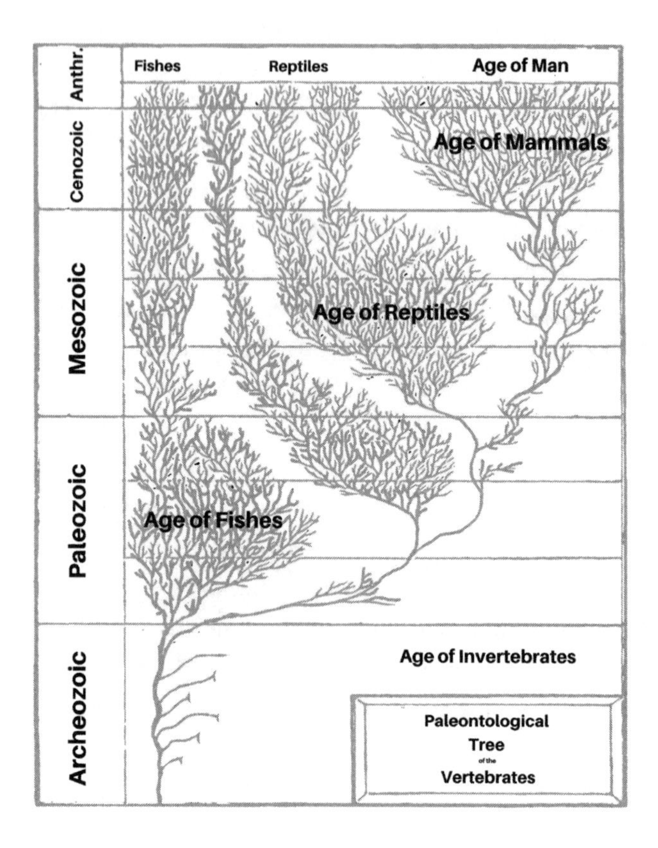

Tree of Vertebrates

Redrawn from Ernst Haeckel's *The Evolution of Man*, 1910.

evolution's tactics. Nonpoisonous snakes copied the vibrant colors of poisonous varieties to dupe predators. The race was on. It continues to this day and will continue as long as species interact with and depend upon one another. Constant change is the way species survive in a complex system. Evolution's endless chase follows the astute observation of the Red Queen, who admonished Alice in Lewis Carroll's *Through the Looking Glass* that "it takes all the running you can do, to keep in the same place."

Diversity of survival strategies holds the key to life's persistence today as much as it did through geologic time. In the aftermath of Hurricane Maria, which wiped out homes, power, businesses, and futures for people across the island of Puerto Rico in September, 2017, plants and animals also had to rise up from the wreckage. In the nearby hurricane-battered islands of Turks and Caicos, lizards had to cling to trees and rocks to survive the storm. The ones that survived had bigger sticky toe pads for firmer grips and shorter billowing hind thighs to avoid being whisked away by the wind. Diversity paid off for renewal of

the lizards after the storm. The lizards that led the renewal have higher chances of surviving the next Maria-like hurricane.

In a world where all lizards have exactly the same size toe pads and hind thighs, a hurricane-strength storm could obliterate an entire population from an island. At the scale of the entire planet, collision with an asteroid or a drastic change in the atmosphere would likely spell the end of life with no hope of recovery.

Diversity is the stabilizing salvation. It is the scaffolding that enables life to withstand onslaughts and adapt to change. Another critical feature for persistence of life on Earth to add to circuit breakers and self-regulating mechanisms: options through diversity. As gene banks are the insurance for the world's food crops, diversity of life is the stabilizing insurance for all of nature.

Diversity in human civilization takes different forms than gene banks and species. Humanity's primary advantage lies in its ability to accumulate knowledge about strategies to survive and prosper. Ideas in people work like genes in other animals. Our species does not need to wait for a genetic mutation or natural selection to adapt its food-securing strategies to changing rainfall patterns. We barter and trade, and we take natural selection into our own hands by breeding and editing genes. Humans can survive a storm with safe buildings and advance warning, rather than big toes and short legs.

The common thread between life on Earth and human civilization lies in the complexity of both systems. In an unpredictable complex system, whether modern society or a forest, diversity serves as a stabilizing force. In evolution, diversity of genes and species provides the leverage to adapt and survive. In human civilization, diversity of ideas, specializations of knowledge, institutions, languages, and cultures serves the same function. The options-open tactic is not lost on investors and engineers, who put the strategy into practice every day to survive through uncertainty.

NATURE'S STRATEGY IN THE BOARDROOM

Even the smartest investors can't predict where to put their bets to maximize their profits. The economy is too complex. By trial and error, financial planners and investors unwittingly learned that nature's strategy yields the best results for their bottom lines. Efficiency, in the form of maximum return for minimum investment, is not always the guiding principle. A long-term investor values stability and spreads the risk across investments to yield minimal fluctuations in returns. "Diversified bet-hedging" was the brainchild of economics student Harry Markowitz in 1952. Four decades later, Markowitz won a Nobel Prize for his lifelong work on portfolio management for individual investors.

A portfolio should include a diversity of investments. Some will go bust and some will reap profit. But even beyond the well-known principle of diverse investments, clever investors strategize to keep diversity in the kinds of investments they select. They should not all fall or all rise simultaneously. In the words of Markowitz, "A portfolio with sixty different railway securities, for example, would not be as well diversified as the same size portfolio with some railroad, some public utility, mining, various sorts of manufacturing etc. The reason is that it is generally more likely for firms within the same industry to do poorly at the same time than for firms in dissimilar industries." Portfolio diversity keeps options open for an unpredictable shift in the market. The stakes are not as high as devastation from an asteroid colliding with the Earth or a lethal pathogen infecting a population, but the principle is the same. Diversity keeps options open in a risky world.

Engineers who design bridges, airplanes, and cars face the same problem as profit-seeking investors and evolution of life on Earth. They cannot know precisely when a part will fail or an unforeseen flood or some other calamity will overwhelm their design specifications. Engineers do not share the potpourri of options that evolution has constructed over millions of years. They cannot manufacture millions of parts, with another part stepping in when one fails.

Mathematician and Hungarian émigré to the United States, John von Neumann, hit upon a solution to the engineers' problem in the midst of the Cold War. Vacuum tubes in early computers often failed. The United States lagged behind the Soviet Union in the reliability of its electronic equipment. Von Neumann proved that multiple computers, though seemingly inefficient and wasteful, actually enhanced reliability. "The basic idea in this procedure is very simple. Instead of running the incoming data into a single machine, the same information is simultaneously fed into a number of identical machines, and the result that comes out of a majority of these machines is assumed to be true." The idea gave birth to the bedrock engineering principle of redundancy. Multiple parts for the same function reduce the chances of total failure. If one part has a one in ten chance of failure, the likelihood that two parts will both fail drops to one in a hundred. No one today would board a commercial airliner without redundant engines, insurance in case one fails.

Redundant parts decrease the chance of catastrophic failure but are far from fail-safe. If identical parts face the same problem, they will all fail. That's what almost happened when British Airways Flight 009 took off from Kuala Lumpur, Malaysia, on June 24, 1982, for an overnight flight headed to Perth, Australia. "Good evening ladies and gentlemen," came the announcement from the cockpit a few hours into the flight. "This is your captain speaking. We have a small problem." The smell of sulfur began to permeate the cabin as the plane cruised 37,000 feet above sea level. Smoke seeped into the flight deck. Passengers saw from their windows an eerie blue glow around the engine housings. Then flames shot from the engines and filled the sky.

Engine failure is reason for concern but far from fatal. Engineers had wisely designed the aircraft with the dire consequences of a failing engine in mind. The aircraft had four engines. In the unlikely event that one failed, three were left to do the job. The aircraft could even fly with only one functioning engine.

Within minutes of the shooting flames, one engine sputtered to a stop. Then another and another and another. Captain Moody announced to the 247 petrified passengers, "All four engines have stopped. We are all doing our damnedest to get them going again."

The temperature in the cabin soared and oxygen became scarce. Masks dropped from the ceiling, but some did not work. The captain, in hopes of preventing passengers from dying of oxygen starvation, sent the powerless, silent plane into a six thousand-foot nosedive in the span of one minute. As the passengers were praying, comforting one another, and writing notes to loved ones, all four engines roared back to life. The pilots landed the plane safely in Jakarta, though one engine failed again and they couldn't see through the sandblasted windshield.

Two days later investigators confirmed what had happened. The plane had flown through a cloud of volcanic dust spewed from Mt. Galunggung one hundred and ten miles southeast of Jakarta, Indonesia. Tiny particles had ground away the tips of the turbine blades and peppered the windshield. Ash had clogged the engines, which blew free only when the plane descended into clean, denser air. The "Jakarta incident" became legend in aviation history and the crew were hailed as heroes. Planes now have monitoring devices for volcanic ash.

In the decades following von Neumann's radical insight, engineers further refined his ideas. They needed a way around cases like British Airways Flight 009 when redundant parts all suffered the same problem. "Design diversity" uses multiple parts for the same function, with each part designed slightly differently. The idea is that parts will fail at different times and in different ways, the same logic as Markowitz's diversified portfolio in railroad, public utilities, mines, and manufacturing. Maybe different teams independently design the same part, or maybe instructions to teams vary slightly for parts that perform the same function. Redundant but slightly diverse parts provide some degree of insurance in a world ridden with potential failures, like the orchid that relies on 21 different species of moths and 24 species of butterflies to carry its pollen rather than any one alone. Over the decades, engineers' strategies have become more like nature as civilization learns to design its machines with uncertainty in mind.

Civilization's experience with the uber-complexity of an interdependent city-world covers only a flash in time. Experience is shallow. Like any complex system, civilization's persistence rests on the ability to right itself from instability activated by an outside force, perhaps a hurricane or earthquake, or an internal influence from political upheaval, bloated and corrupt bureaucracy, or a financial circuit-breaker that doesn't trigger in time. Despite track records of investors and engineers whose tactics resemble those of life on Earth, modern civilization is running roughshod over nature's proven strategy in at least three other ways: the words we speak, the food we eat, and the microbes that live in our guts.

UNSPOKEN WORDS

My mind often wanders to the place in the heart of India, where tigers and people live together in a landscape peppered with forests, villages, and rice fields, where my students and I travel to learn about the realities of the place. The bus stand in the village where we stay bustles with activity when the rickety red bus pulls up on the dusty road. Men, women, and children pile into the wooden seats and scramble on top of the bus, carrying baskets of vegetables, headloads of wood, shopping bags, and propane cylinders. A gaggle of languages rises from the roadside crowd as people hustle to board. Most speak Hindi, the mainstream language of the region. Tourists and visitors from the cities converse in English. A few might speak one of the many local languages and dialects of the local tribal peoples, Gondi spoken by the Gonds and Chhattisgarhi spoken by the Baigas.

Gonds, the ancient tribal group spread across central India, number almost three million. But today only one in four speak the Gondi language. Those who still live in the most remote forests and learn the secrets of subsisting from plants and animals retain the traditional language. For those families who connect with the modern world and travel to town in rickety buses to shop and study, the dominant state language seeps in over a few generations. In one generation, parents might speak the language at home, but children use the state's official language in school and when interacting with the outside world. The next generation might not learn Gondi at all. Only the elders maintain a memory of the traditional language. With the last breath of the last elder, the language dies. In the span of a few generations, languages can go extinct just as the dodo bird and passenger pigeon perished at the hands of colonizers.

The rich tapestry of cultures and languages grew with the knowledge of plants, animals, memories, beliefs, and survival strategies in each new place where people ventured. Today, approximately seven thousand languages are still spoken somewhere in the world. But the numbers are deceiving. The mother tongue for half of all people is one of only two dozen languages. The top ten: Mandarin Chinese, Spanish, English, Hindi, Portuguese, Bengali,

Russian, Japanese, Javanese, and German. The remaining thousands of languages are spoken by small groups of mostly indigenous people in remote places. Diversity of languages clusters in places with high diversity of plants and animals.

In India, for example, more than four hundred languages and many more dialects compose a medley of cultures and languages from tribal, Dravidian, and Indo-European roots. On the other side of the world, Native American languages and those brought by immigrants account for more than three hundred languages spoken in the United States. Most remaining languages throughout the world face danger of extinction as grandparents and great-grandparents are the last repositories of knowledge.

Like species in the long history of life on Earth, new languages formed and others went extinct in fits and starts throughout human history. In Eurasia, populations grew and expanded outward from Vavilov's center of origin with the advent of farming millennia ago. People splintered into fragmented groups and languages diverged, giving rise to the vast family of Indo-European languages that include Hindi, Spanish, English, German, French, and Persian. The expansion displaced other languages, long lost to history, of small bands of hunter-gatherers. Similarly, waves of Bantu expansion throughout Africa displaced many languages and gave rise to hundreds of new ones as people dispersed into different places.

More recent extermination of languages in the last few centuries is far more drastic. Half of all known languages have died in the past five hundred years. When Columbus arrived on America's shores, people living in what became the United States collectively spoke hundreds of languages. The number has dwindled by almost half, and most of those are barely surviving through a small number of elder speakers. A similar story holds for Aboriginal languages of Australia. Genocide, disease, and cruel measures on the part of European conquerors to forcibly remove children from parents and forbid them to speak their native languages stamped out hundreds of local tongues in previous centuries.

Today, the loss of languages continues, although the forces are not as tragically deliberate. In the modern ultraconnected world, survival of languages tuned to local cultures and spoken by small groups of people swims against a strong current of mass communication and globally shared ideas. Many more languages spanning across all continents are likely to go extinct within decades, as children don't learn from their parents and communities. One, the Majhi language spoken in a tribal area in the northeastern hills of India, died when the last speaker, octogenarian Thak Bahadu, passed away. Newspapers reported the deaths of the man and the language on July 22, 2016. Charles Darwin saw the parallel. "A language, like a species, when once extinct, never . . . reappears," he wrote.

If one goes by the Old Testament, one can conclude that the plethora of languages in the world is a curse and a drag on humanity. God, dissatisfied with human hubris in building a tower in Babylon tall enough to reach heaven, punishes humanity with a multitude of languages. Without a common way to communicate, people would no longer be able to understand one another and could not build the tower. So goes the biblical origin story of why so many languages arose in the world. The story has a valid point. If small groups of people can only communicate among themselves, the benefits of a connected world to share ideas and spread technologies suffer. A common language is efficient. But the other side of the story is equally valid. The multitude of languages, rather what languages represent, brings diversity into humanity's tool kit of survival strategies.

Language encapsulates much more than words to communicate about things, places, and feelings. Language opens windows into culture, belief systems, and generations of accumulated knowledge about plants and animals that supply food, medicine, and spiritual renewal. Which plants are good to eat and which are poisonous? Which bark, roots, and leaves provide medicines that can cure an ailment? When is it safe to catch fish without depleting the supply, as only intricate knowledge of the breeding habits of different fish species can reveal? How does a community make collective decisions? What is our place in the world among the vastness of all life and time? When children don't learn the vocabulary to keep knowledge alive, the death of a language can extinguish practical skills and deep wisdom acquired from long experience with a place.

In the words of the linguist and storyteller Basil Johnston, one of the last native Anishnaabe speakers of the Ojibway First Nations in what is now Ontario, Canada, when a language dies, native peoples "can no longer understand the ideas, concepts, insights, attitudes, rituals, ceremonies, institutions brought into being by their ancestors . . . no longer are the wolf, the bear and the caribou elder brothers but beasts, resources to be killed and sold." Language no longer communicates a way of life passed down through generations. The Anishnaabe word *w'daeb-awae*, as Johnston describes, roughly translates to mean that someone is telling the truth. But the true meaning is more nuanced: "The speaker casts his words and his voice only as far as his vocabulary and his perception will enable him." The difference reveals an underlying philosophy. "The best a speaker could achieve and a listener expect was the highest degree of accuracy" rather than the existence of an "absolute truth," Johnston explains, revealing that a person's conclusions can change over time counter to a fixed clockwork view of the world.

An Aboriginal community living on the western edge of Cape York in northern Australia does not use words for right and left to communicate positions of people or objects. Speakers orient locations in space by the path of the sun. The location of an object is east, west, north, or south relative to another. "The boy standing to the south of Mary is my brother," they might say in the Kuuk Thaayorre language. To speak, one must stay oriented in space, a useful skill in the flat expanse of the peninsula. With the cognitive tool to keep track of one's location honed through language, even a child can navigate through the landscape better than an adult speaker of a language based on "next to," "behind," or other relative positions.

In the modern world, profound knowledge about places, prowess to navigate without maps, and appreciation for the transiency of perceived truths might seem irrelevant. But recall that many of today's most essential medications arose from traditional knowledge about the curative power of plants. Ancient Egyptians and Greeks knew that the bark of a willow tree relieves pain and reduces fever. Thousands of years later, European pharmacists derived aspirin from the same chemical compound. Similarly, Chinese knew about the antimalarial properties of leaves from a wormwood plant more than two thousand years ago. Today, the compound is the most effective antimalarial medicine available. European explorers in South America saw Amazonian Indians shoot poisonous arrows concocted from plants, which eventually provided the knowledge to produce muscle relaxants in surgery, antidepressants, and treatments for Parkinson's disease.

No one can say what treatments unknown to modern civilization reside in traditional knowledge. No one can foretell which belief systems or cognitive options might prove most fitting as the liabilities from our technology-driven world come to the fore. But one can be sure that if that knowledge disappears, we will never know the answer. Humanity's options diminish as each language, and the knowledge that goes with it, fades into forgotten history.

Deep knowledge of survival through variable climate in the past might help persist through future uncertainty. The Inuvialuit in Sachs Harbor far north in the Canadian Arctic hunt musk-ox, polar bears, foxes, caribou, and seals in winter when hunters can safely traverse the ice. They have long lived with vagaries of the weather. Flexibility is their chief strategy. The Inuvialuit's detailed hunting plan is never the same in any two years. It depends on when the sea ice freezes and places hunters can be sure to reach. Their words specify subtle differences in the types of fractures in the ice. Especially with freezing later in the fall and melting earlier in the spring as climate change hits the high latitudes, "you really have to watch" in the words of one Inuvialuit. Contrast the Inuvialuit's close observations with the official government's fixed delineation of the start and end dates of the polar bear sport hunting season, without regard for the actual condition of the ice.

The modern, interconnected world smooths over the rugged richness of cultures, knowledge, and belief systems. Vanishing languages are a warning sign that knowledge rooted in different places and acquired over millennia of experience is melding into a homogenous stew. The trend runs counter to the basic principle that diversity maintains options for uncertain times. And the loss of diversity in language and culture goes hand in hand with the massive shift in the foods people around the world grow in their fields, purchase from stores, cook in their pots, and eat for dinner.

ON THE PLATE

Norman Borlaug, father of one of modern civilization's most incredible achievements, was a zealot with ambition to feed the world. His painstaking work to breed high-yielding wheat varieties boosted the amount of cereals that farmers could produce from the same land. It set off the world-changing Green Revolution. Quantities of rice, wheat, and maize soared in the second half of the twentieth century with fossil fuel–powered machinery, fertilizer, irrigation, and high-yielding varieties. Borlaug could be proud of the revolution's achievements. It lifted many farmers throughout Asia and the Americas out of poverty; ushered in an era of cheap food; and relegated periodic famines in the Indian subcontinent to horrible memories.

As much as Borlaug was proud of these achievements, the Minnesotan farmer was also a realist. He knew that the Green Revolution was not a panacea. It was, as he remarked while accepting the Nobel Peace Prize in 1970, "a temporary solution, a breathing space, in man's war against hunger and deprivation."

As options foreclose when languages vanish with elders, options to fight the war against hunger and deprivation narrow as edible plants and animals disappear from farmers' fields, crop breeders' genetic stocks, and markets where people buy their food. The phenomenal diversity of life on Earth includes at least 250,000 plants, a number that is probably a gross underestimate. A small subset, about 30,000, are edible for humans. At some point in history, people have collected and eaten an even smaller subset of these edible plants, about 7,000.

Expeditions to enhance the basket of options from distant shores have been a part of history since people could set sail across the seas. The earliest known botanical expedition dates from ancient Egypt, when Queen Hatshepsut dispatched five ships across the Red Sea to the Land of Punt to gather valuable spices and myrrh trees for their aromatic resin. Many centuries later, following Columbus's venture across the Atlantic, maize traveled to Africa, potatoes to Europe, and sugar to the Americas, among thousands of other plants. Diets around the world changed forever. The value of plant options was even clear to the president of the United States. Thomas Jefferson, the third president, stated: "The greatest service which can be rendered any country is to add a useful plant to its culture."

The efficiency paradigm of modern civilization has shrunk the basket of options. Borlaug's Green Revolution honed in on the cereals most amenable to technologies that could boost yields and produce the greatest quantities. The decision was logical. People were starving. Famines loomed on the horizon. But like so many decisions in a complex world, the by-products are hard to foresee and difficult to remedy afterward.

Today, a tiny fraction of the seven thousand plants that have fed humanity are part of the modern diet. Half of humanity's plant-derived calories come from only three plants: wheat, rice, and maize. An additional six add another quarter: sorghum, millet, potatoes, sweet potatoes, soybean, and sugar, and a handful of animals bring calories into the human diet: cows, pigs, goats, sheep, chickens, and fish. The basket shrinks even further when one considers the variation within each species on this short list. Types of rice number in the tens of thousands, for example. Most are traditional varieties that farmers continually adapt to their local soil and rainfall by saving the seeds of the most successful offspring from the year before. The high-yielding Green Revolution rice ran roughshod over lower-yielding but better-adapted varieties than the ones bred by official plant breeders. When farmers adopt the high-yielding varieties and abandon the locally adapted ones, genetic diversity with the potential for future breeding disappears, as did the Majhi language that blinked out with Thak Bahadu's last breath. Apples followed the same trajectory. More than eight thousand varieties of apples grew in the United States at the turn of the nineteenth century. Only a few hundred remain today.

The race is on to capture the diversity of landraces in gene banks before it's too late. Even with heroic efforts, the prospect of capturing even a fraction is slim. Even less possible to store in a vault is the living, dynamic evolution of varieties that continually adapt to changing climate and pests in different places.

Technologies of the Green Revolution have not yet reached the village where I live in the heart of India. Animals pull plows and people stoop to plant rice by hand when the monsoon brings rain. The villagers get abysmally low return for their hard efforts. But one by-product of the revolution that boosted the country's economy hangs on display in the small kiosk near the bus stand. Strips of aluminum packets of cookies, crackers, potato chips, and candies glimmer with enticement. Children run up to the kiosk, offering a few rupees and joining the massive change in diets that is sweeping across the world.

Maybe the culprit is mass marketing. Maybe it's the convenience of processed foods that stay on the shelf without spoiling. Maybe it's the abundance of corn that makes fructose-sweetened drinks so cheap that people can consume them like water. Maybe it's people's evolutionary attraction to sugar. Or maybe sugary, fatty, and processed foods just seem delicious. Whatever the reason, people have become more similar in the types of foods they eat regardless of where in the world they live or how much they earn. Sadly for the health of the world's population, cheap and efficiently produced oils, sweeteners, and cereals for animal feed define the model. The unmistakable trend is toward more processed, sugary, and fatty foods. The allure, affordability, and wide availability extends from the United States, Mexico, Egypt, South Africa, and around the world to my Indian village.

No country seems immune to the sharp uptick in obesity. High-sugar and high-fat diets bear some of the blame. Repercussions are enormous for a lifetime of health woes from diabetes and heart disease. Modern civilization solved one problem of too little food and created another with too much of the wrong kinds of food.

As diets become more similar around the world, options to learn from alternatives are fading. Okinawans live on the Ryukyu Islands in the southernmost prefecture in Japan. They have the enviable distinction as the longest-living people in the world, with the highest rate of centenarians and low risks of diabetes, cancer, heart disease, and age-related ailments. Sweet potatoes are a staple in their diets. They eat Chinese okra, pumpkin, seaweed, bitter melon and a diversity of other green and yellow vegetables, tofu, and small amounts of fish and pork. Calorie intake is low, like the Biospherians during their two-year lock-in. A search on the Okinawan diet yields many studies linking their lifestyle and diet to their disease-free longevity. The island dwellers have the secret that could save the rest of the world from a diet-induced path to destruction.

By the year 2000, another reality had taken hold in Okinawa. Life expectancy of young Okinawans began to drop, particularly for young men who avoided the traditional dishes. A 2004 headline revealed the story: "Love of U.S. Food Shortening Okinawans' Lives/Life Expectancy Among Islands' Young Men Takes a Big Dive." An American base on the island, established after World War II, spread American habits—cars, suburban malls, and fast food. The younger generation couldn't resist. "There are a lot of burger fans like me in my generation," one young man told the reporter. The older generation hangs on to the traditional diet. But like languages, habits and knowledge fade with elders' passing. Were it not for the sleuthing of scientists to reveal the Okinawan enigma and distinguish diet from genetics as the reason for their longevity, the worldwide homogenization of diets would have swallowed the secret. Options offered by the diversity of diets around the world would have diminished.

As Borlaug warned, the success of cheap food has indeed proven itself to be a double-edged, temporary solution.

COMPLEXITY IN THE GUT

Diversity's stabilizing role applies not just to languages, landraces, and diets. Everyone's digestive tract hosts a bonanza of diversity in the form of tiny microbes. One hundred trillion cells weighing over three pounds reside along twenty feet of the human gut. Microbial species in the gut collectively harbor ten million genes, many times more than the human genome. We could not live without these microbes, just as they could not survive without us. Who they all are and what they do in our guts lie on the frontier of knowledge that is opening with novel technologies to unravel their genes. The knowledge could "become the first great breakthrough of twenty-first century medicine," reflects a gut microbiologist who urges us all to take care of our microbiomes.

Counter to the notion that microbes are disease-causing germs, the vast majority work in our favor. Even Louis Pasteur, who promoted the germ theory of disease and saved millions of lives with his pioneering work on vaccines, wrote in 1885 that animals totally deprived of "common microorganisms" would not be able to survive. In the gut, microbes break down the food we eat and supply us with energy, vitamins, and enzymes. The good ones fight back against bad, harmful pathogens. In turn for these services, gut microbes get energy and nutrients for their own survival from their hosts' food.

Each type of microbe has its role. Those grouped as Firmicutes extract energy from fat. Bacteriodetes break down and ferment starchy carbohydrates. Much is left to learn, but our dependence on the vast diversity of life in our guts is clear. As Jeffrey Gordon, father of the field that revolutionized understanding of the gut microbiome, reminds us: "Microbial communities are wondrous things; they form; they are able to persist; they adapt nimbly to changing circumstances; they are resilient; and the . . . interactions are complex and dynamic. All of us . . . have a sense of awe, coupled with a feeling of humility, when exploring the terra incognita that is the gut microbiome."

Like any living system, the community of microbes adapts and changes with its environment. A community gets its start as a baby exits its mother's womb. Diversity of microbes rises as babies gets older and the community changes during the first few years of life. In adulthood, the types of microbes living in our guts depend on where we live and what we eat. Fecal samples from Venezuelan Amerindians in two villages; four rural villages in Malawi where maize dominates people's diets; and Americans from Boulder, Philadelphia, and St. Louis show vast differences in microbial communities. Gut diversity of Venezuelans and Malawi were similar to each other and both were more diverse than the American gut.

Low gut diversity and a higher proportion of Firmicutes compared with Bacteriodetes seem related to the Western lifestyle and highly processed diet. Unfortunately for the young Okinawans who took a liking to the American diet, these features of the American gut also appear to go hand in hand with the chronic afflictions of obesity, digestive diseases, and cancer. The stabilizing force of diversity in the complex system of microbial communities isn't surprising. It stands to reason that microscopic diversity mirrors the essential role of diversity in macroscopic forms of life. Similar forces are at play. Microbes compete and cooperate in the Red Queen's chase just like cheetahs and gazelles or bees and flowers. A varied diet of proteins, carbohydrates, and fats offers a buffet that maintains a diverse population of microbes in the gut, which in turn provides a range of vitamins and enzymes to keep people healthy.

Just as Borlaug was not thinking about diversity of crop varieties and their wild relatives when he set out to alleviate hunger by increasing cereal yields, the value of gut diversity was not on the mind of German scientist Paul Ehrlich when he set out to find a cure for a dreaded disease at the beginning of the twentieth century. Decades later, the downside of society's profligate use of the life-saving discovery are rising to the fore.

Ehrlich was a fan of Sir Arthur Conan Doyle's Sherlock Holmes mystery stories. The thrill of tracking down clues might have inspired his search for a "magic bullet," as he called it. He was looking for a cure for syphilis, a disease common in Europe at the time caused by the newly discovered, sexually transmitted twisted spirochete bacteria. The clue came from his observation that synthetic dyes stained some microbes and not others, meaning a toxic substance could target a specific microbe like a bullet.

With syphilis-infected rabbits as guinea pigs, Ehrlich and his coworkers tested substance after substance by injecting the rabbits with different forms of arsenic-containing powder. Finally, in 1907 on the 606th try, they hit upon the right combination. The drug Salvarsan, also called 606, achieved great success in curing the disease, but the treatment was long and painful. The search for a "magic bullet" was still on.

The era of antibiotics inched closer when Scottish physician Alexander Fleming serendipitously discovered penicillin on September 3, 1928. Fleming returned from a family vacation to his untidy lab. He found spots of a blue-green mold, the type that grows on bread, in his culture plates. The mold could have flown in the window or come up through the floor. The "mould juice," as Fleming named it, had killed the bacteria that he was trying to grow.

More than a decade passed before an Australian, Howard Florey, and a German refugee, Ernst Chain, started to experiment with the mold and turn it into the wonder drug that it became. Penicillin saved many soldiers in World War II from infections and afterward many millions of people from gangrene, pneumonia, and tuberculosis, among many other bacterial ailments. It replaced Salvarsan to treat syphilis and seemed to be Ehrlich's "magic bullet." When the three—Fleming, Florey, and Chain—received the Nobel Prize in 1945, the modest Fleming remarked, "It was destiny which contaminated my culture plate."

Antibiotics rank as one of the greatest life-saving achievements of the twentieth century. I, for one, would probably not be here were it not for antibiotics to cure a bout of pneumonia in my twenties. As with most world-changing achievements, collateral outcomes come to the fore over time. The widespread use of antibiotics is no different. One well-known problem is the arms race between new antibiotics and antibiotic-resistant bacteria spurred by overuse of the miracle drugs in treating livestock and people. Another resembles the loss of languages to efficiency of communication and the genetic erosion of our food supply with high-yield crops to feed modern civilization.

"Magic bullet" antibiotics do not just target bad bacteria that cause syphilis, pneumonia, and other life-threatening diseases. They also kill good microbes in the gut. The effect on the microbiome can persist for years. Ehrlich and Fleming had no way of knowing that twenty-first-century discoveries would reveal that their miracle cures did indeed save lives, but they also contribute to diminished diversity of gut microbes tied to the crippling chronic ailments of the present day.

Germ-killing antibiotics or healthy gut microbes? Common or many languages? High-yield cereals or diverse crop varieties? One does not negate the benefits of the other. But the two uncomfortably coexist in the modern world. High-diversity options have a tendency to fall by the wayside in our interconnected world. Perhaps they just don't seem relevant from a short time horizon, or maybe the costs seem to outweigh the benefits. In our modern world the scales are tipped toward the short-term, low-diversity solutions at the expense of longer-term diversity. But arrays of languages, species we depend on for food, and gut microbes might be well worth the price to provide options in our uncertain, modern civilization.

Life on Earth speaks from long experience with its second piece of advice for modern civilization. Value your diversity: in your guts, your institutions, your conceptions of knowledge, your cultures, and in the plants and animals that underpin your survival. Diversity has provided a portfolio of options for life to persist for billions of years. It brings insurance to recover after a fall. Investors discovered over time the value of diversity for their bottom lines. Engineers learned that routinely building redundant and diverse parts into their designs is a matter of safety.

Leaving aside the short-term interests of investors and engineers, the tendency of the modern world in many aspects of everyday life is to squash diversity in the interest of efficiency. If everyone spoke the same language, ate the same food, and grew the same varieties, efficiency would prevail, but civilization would lose reservoirs of options for changing times.

Diversity seems messy and is not always easy to maintain in our human-constructed world. To keep it requires thought, planning, organization, and money. A heedless thrust to short-term efficiency will run roughshod over diversity. But don't be fooled.

You do not need to make a false choice between diversity and efficiency. Stock your gene banks, protect your remaining wild places, bolster your efforts to keep languages alive, and value your lifestyles that keep diversity in all aspects of civilization from your ideas to your belly. Resist the temptation to skimp on diversity in the name of expediency. Be wary of "magic bullets" that diminish your options for the future.

4

MIND THE NET

Defend Against Cascading Failure

IF there is a single day when it all began, that day would be April 26, 1956. From the crowded, grimy waterfront of Newark, New Jersey, a converted World War II tanker named *Ideal X* set off to haul its cargo down the Atlantic coast to the Gulf of Mexico. The tanker was headed to the Texas port of Houston. The extraordinary voyage was a long time in the making. And it changed the path of civilization as much as wind-powered ships, rail lines, the internet, or any of the other revolutionary leaps in technology that connect people and economies across the world.

Two decades earlier, an ordinary trucker from North Carolina stood idly on the shore in Hoboken, New Jersey, killing time while waiting for his truck's cargo to be loaded onto a ship so he could head home. At the time, young Malcolm McLean ran a trucking company that he had started three years out of high school. McLean and his siblings picked up extra dollars hauling empty tobacco barrels in an old trailer during their high school years. Trucking cargo up and down the coast seemed a logical next step for the family business.

As McLean waited and watched the longshoremen load and unload cargo from the ships, an idea struck: bypass the cumbersome steps of unloading cargo from the truck and then reloading onto the ship. Instead, just put the cargo in giant metal shoebox-type containers right at the origin. No need to unpack the cargo to load boxes from trucks or trains directly onto the ship. Repeat at the other end to deliver the container to its destination. All the scheme required was standardized containers, trucks, trains, cranes, ships, and storage facilities.

The process of shipping goods had changed little since ancient times. Longshoremen loaded cargo onto a ship at one end and another group of longshoremen unloaded it at the other end. Carts, trains, or trucks carried odd-sized sacks and boxes of cargo to their destinations. The time-consuming and costly process was the loading and unloading, not the actual shipping, and all those changes of hands led to pilfering on either end. McLean's genius was to integrate the trucking and shipping. The process eliminated the bottleneck of loading and unloading in ports, drastically cut costs, and eventually revolutionized global trade.

Fifty-eight containers rode on specially rigged decks in *Ideal X* as McLean tested the smooth transfer of cargo from truck to ship to truck on the receiving end. The "ignored pioneer," remembered in his obituary as "one of the few men who changed the world," amassed a fortune from his simple idea.

Economists debate whether container technology or trade policies were responsible for soaring trade volumes that connected the world in the decades since *Ideal X*. Certainly, the two go hand in hand. Just as certainly, twentieth-century globalization was not the first time the world appeared smaller as far-flung economies and cultures grew more intertwined. Many millions of years ago, our ancestors ventured from the grasslands of Africa into Europe, then Asia, and eventually the New World. Then agriculture allowed people to settle in one place, make pottery and other specialized goods, and trade. Our species, unlike others, was no longer restricted to places where it could find food and raise its young. For centuries, trade routes along the Silk Road mixed goods, cultures, and religions from China to the Mediterranean coast. Globalization leaped forward again with the massive exchange of plants, animals, and pathogens following Columbus's foray to the New World. Corn and potatoes, both New World crops, became staples in the diets of cultures around the world. Again in the early twentieth century, before World War I, the Panama and Suez Canals opened routes for a global economy. In more recent times, with shipping containers and goods exchanged throughout the world, the movement of people, ideas, and products wove more threads into the vast net that now connects nearly all places and all people.

McLean's legacy is on display at ports and distribution centers around the world. Every day, trucks, trains, and planes deliver more than twelve million pounds of fruits, vegetables, meat, and fish to massive, sprawling warehouses on the tip of a peninsula at the southern end of the Bronx, New York. Food from fifty-five different countries and forty-nine states flows into Hunt's Point, the world's largest food distribution center. Trucks line up at the loading docks and deliver their cargo to more than forty thousand convenience stores, supermarkets, and restaurants scattered throughout the city. From there, millions of people lug their groceries and carry-out orders to their apartments and homes. The maze of connections links the kitchens of New Yorkers with faraway orchards that produce fruit, fields that grow vegetables, farms that raise animals for meat and milk, and seafaring vessels that catch fish. The city simply could not function without the vast food distribution network.

Cities and towns all over the world have similar, though less colossal, networks that link places where farmers produce food with the people who consume it. Not just food, but energy, water, information, ideas, finances, and people flow through pipes, power lines, highways, airline routes, shipping lanes, word of mouth, and the internet. Networks have been the lifeblood of civilizations dating back to antiquity. In the Incan empire of the high Andes, for example, with no written language or wheels for vehicles, highly trained runners relayed messages along a network of huts connected by mountain paths. In ancient Rome, networks of aqueducts carried water from distant hills to cities and towns.

Networks are the mainstay of today's modern world. Airline networks connect cities across the world. Crisscrossed shipping routes ferry goods from port to port. Financial markets thrive on the flow of money between banks. The internet transmits information from router to router. Power grids transfer electricity along wires and cables into billions of homes and businesses. From buzzing cities to remote villages, nearly every person in the globalized, interconnected world depends on networks. They lie at the heart of the burgeoning material prosperity of the last several decades.

Networks make modern civilization possible. But they also bring risks. When one part breaks, failures can cascade throughout a network and cause catastrophe. One such failure happened on June 7, 2016, in the Gitaru power station in central Kenya. A rogue monkey climbed onto the roof of the power station and tripped a transformer, which in turn triggered a nationwide blackout. The monkey survived, but the blackout severed internet services, left the entire country in the dark, and paralyzed businesses for more than three hours.

Seemingly minor, unpredictable monkey-like disruptions propagate through networks. Examples are endless. The ash cloud from the 2010 explosion of the Eyjafjallajökull Volcano in Iceland stranded travelers and crippled airline networks for many days. Hepatitis, HIV-AIDS, and other blood-borne pathogens spread from blood donors to unlucky recipients until accurate tests and safeguards were in place. The Love Bug worm promulgated similar cascading failures across the internet on May 4, 2000, as people opened the attachment to a message with ILOVEYOU in the subject line, infecting millions of computers around the world.

Life on Earth relies on networks at least as much as modern civilization, with similar benefits and risks. From microscopic veins that distribute water throughout a plant's leaf to neuron-transmitting and blood-carrying networks in animals' brains and livers, evolution has rewarded the life-enabling advantage of networks. At grander scales, food webs transfer energy from the sun to plants and from plants to animals. From the cells of the smallest living creature to the global-scale cycling of nutrients, networks are everywhere in nature.

Like the human-constructed varieties, nature's networks are prone to cascading failure. Disaster could strike if a munching insect severs a vein carrying water through the leaf. Or a food web could come crashing down if bottom-of-the-pyramid plants or insects die off from disease or from lack of water or food. Through long experience with networks, life on Earth has undoubtedly picked up a few tricks to exploit the benefits and downplay the risks. Humanity is unwittingly learning that the network design of tiny veins in a leaf might function better than a seemingly efficient, hub-and-spoke, "small-world" network typical of airline routes, the flow of information in social networks, and many other human-constructed networks.

THE NONRANDOMNESS OF NETWORKS

Networks in the real world come in many patterns. Some are simple physical networks, such as regular, equally spaced grids like city blocks. Some networks have clusters, such as social networks to share information among friends and families, with cliques of people who communicate mostly with each other and a few popular people who communicate across cliques in a hub-and-spoke pattern. Transport networks for water, ranging from tiny plant roots in the soil to large river networks that drain into the sea, follow a branching, treelike pattern.

Life's networks have many varied patterns, but they share some common characteristics. Rarely do they follow a regular, gridded pattern like a crystal of table salt. Nor do they follow a random pattern of nodes connected with equal likelihood to any other node in the network.

Stuart Kauffmann, the brilliant American medical doctor who turned his attention to the study of complex systems, explains random networks with buttons as nodes and thread as links. In a room littered with buttons, pick two at random and connect them with a thread. Repeat the process, connecting two buttons with thread. Eventually, every button will connect to another in a giant cluster. Now lift any button. Any button you pick will pull up the entire knotted web of buttons and threads.

Many networks in the real world actually follow this basic pattern. Epidemics can spread through entire populations, for example, and social networks connect people across the world through mutual acquaintances. But there is a key difference between theoretical random networks and the real world. In a random network, each node is equally likely to be connected to any other node anywhere in the network. That's not how the real world behaves. Neighbors are more likely to be connected to nearby neighbors rather than someone across the world. Or perhaps people cluster according to political views, preferences, upbringing, or some other nonrandom factor.

In network parlance, reality can resemble a "small world," a term popularized by the well-known six degrees of separation in social networks and extended to the internet, power grids, disease spread, terrorist organizations, and other aspects of our connected age by pioneering network scientist Duncan Watts and colleague Steve Strogatz. In small worlds, a few key links serve as hubs that keep every node in the network connected. Besides these few hubs, most nodes tend to be highly clique-ish. The architecture of small-world networks lies between a city block–type, regular lattice, with nodes only connected to local neighbors, and random networks, with distantly connected, unclustered nodes.

The electrical power grid of the western United States, for example, follows the small-world pattern.

Networks of sexual partners might also follow a classic small-world pattern, according to a study in Sweden in the 1990s. Researchers questioned 4,781 Swedes between the ages of 18 and 74 about the number of partners they had during their lifetime. Unsurprisingly, males reported more partners than females. More surprisingly, a few men and women reported hundreds of partners, while most reported a small number. The same patterns emerged from similar surveys in the United States, Britain, and Uganda—a small number of highly promiscuous people and a large number of people with a small number of partners.

In 1999, Hungarian-born network scientist Albert-László Barabási and colleague Réka Alber explained why reality so often follows the few hubs–many clusters pattern. They invoke the "rich get richer" phenomenon. As networks grow, new nodes preferentially connect with nodes that already have many connections. Promiscuous individuals attract more sexual partners than those who are more restrained. Famous artists become more famous, while undiscovered artists sink into oblivion. A new airport will link to a large hub in an airline network rather than another small airport.

Barabási and colleagues came to this simple yet profound conclusion through crawling the World Wide Web, the largest human-created network to date with billions of web pages. They counted how many times each web page links to other web pages. They expected to find a standard bell-shaped curve, similar to the distribution of human heights. In a bell-shaped curve of heights, most people are as tall as the average height, some are taller, and some are shorter. A random network should have a bell-shaped distribution for the number of web pages that link to each page. "Instead," they reported, "we discovered certain nodes that defied explanation, almost as if we had stumbled on a significant number of people who were 100 feet tall." It was the same pattern as the network of Swedish sexual partners, a few nodes with a huge number of connections and most nodes with only a few. They had uncovered a small world in the extreme.

Sexual partners, web links, and other small-world networks share a common trait: most nodes connect to each other only through a major hub. Many networks in the modern, globalized world share this characteristic to varying degrees, including the spread of ideas and ideologies, trade partners, hub-and-spoke transport networks, and financial markets.

The distinction between random and small world might not have much consequence on an ordinary day. But it makes all the difference when catastrophe strikes. A network's architecture can control whether a country plunges into darkness from a monkey jumping on a transformer, whether a drought in one part of the world sets off food riots in another, whether one sick person triggers an epidemic, or whether a leaf dies when an insect severs the flow of water through its veins.

Networks, whether in nature or the human-constructed world, suffer from the same contradictions. They bring great benefits and also great risks. They protect and they expose. And they are a fact of life in modern civilization. Small-world networks, which grow in a globalized, interconnected world, bring particular trade-offs.

The characteristic feature of small-world networks—a few hubs and many small clusters—makes them vulnerable. The monkey that jumped on a transformer happened to hit a hub in Kenya's power distribution network and cause a nationwide blackout. Disaster can ensue with the loss of a major hub. All the connections linked to the hubs get severed. Cascading failure ripples and propagates throughout the network.

Human-constructed networks become small worlds because the architecture emerges from sensible decisions. Buy from the largest supplier with the most number of other customers; attach a new airport to a major airline hub so travelers do not need to make too many connections; click on the web page with the most number of prior hits. Small-world networks emerge from these commonsense decisions. They are efficient, with a small number of links to transfer goods across an entire network. And if one of the many non-hub nodes gets taken out by a monkey or a delayed airliner, the whole network still functions. But these efficient, sensible decisions that result in small-world networks can bring enormous trouble when an unexpected disruption hits a hub.

Car makers and computer manufacturers experienced the fragility of global trade networks in 2011. Factories in Thailand were supplying about half of the world's hard disk drives at the time. When unusually strong monsoon rains hit in October, neck-deep floodwater inundated many of the factories. People in boats tried to salvage the hard disk drives, but their efforts could not stop the cascading, global effect on the electronics and auto industries, which had already placed orders for the drives. Shortages and higher prices lasted for many months. Thailand is the hub because it supplies inexpensive disk drives to many industries around the world. The supply is reliable most of the time. But the flood-induced fragility of that hub-heavy supply chain proved costly.

Whether networks of social contacts follow a random or small-world pattern can be the difference between a small, isolated disease outbreak and an epidemic. Take the case of typhoid-causing bacteria. Bacteria spread typhoid through feces of infected people that comes in contact with food or water. If an infected person contaminates food or drinking water that not many other people eat or drink, the disease would spread slowly through the population, perhaps dying out if infected people recover or die before passing on the bacteria to others.

The most famous "super-spreader" of all time, Typhoid Mary, proves the power of hubs in a small-world network. Mary Mallon was an Irish woman living in New York in the early 1890s. She made her living by cooking for wealthy New York families. Mary had an uncommon ability to carry the bacteria but not succumb to the infection. Everywhere she cooked, within weeks the families fell ill with typhoid. The bacteria from her feces must have made its way into the dishes she prepared, likely through unclean hands. Eventually, the authorities quarantined Mary and she spent the rest of her life in isolation. Mary was a hub.

With air travel in more recent times, super-spreaders can proliferate deadly diseases even farther and faster. A physician from Guangdong Province in China was the most significant of four super-spreaders in the 2003 outbreak of Severe Acute Respiratory Syndrome, or SARS, a disease caused by a coronavirus similar to Covid-19. The doctor had contracted the disease from patients before attending a family wedding in Hong Kong. From the ninth floor of the Metropole Hotel, family and friends attending the wedding carried SARS to Singapore, Vietnam, Taiwan, and Canada. They exposed hundreds of people on their flights as the virus spread through the air when infected wedding guests sneezed, coughed, or talked. Within a few months, SARS spread to thirty countries before finally subsiding.

From trade networks to social contacts, the tendency of the modern world is to construct small-world networks. These networks have enormous benefits for efficient and functional societies. But they leave civilization vulnerable and exposed to cascading failures emanating from a broken hub. Small-world networks pose a contradiction: efficiency under normal conditions versus fragility when an unexpected shock inevitably hits a hub.

HOW A LEAF RESOLVES THE PARADOX

Life on Earth's networks face the same trade-offs as human-constructed ones. Should networks be efficient but fragile, or should they be resilient at the expense of efficiency? How much investment in a less fragile, more resilient network is worth the cost to prepare for a time when calamity strikes? Nature has evolved over the last four billion years, time enough to experiment with different network architectures to reconcile these trade-offs. With high-resolution images of microscopic veins on leaves, some clues to nature's resolutions are coming to light.

The intricate veins in leaves are elegant two-way transport networks. In one direction, veins distribute water throughout the leaf. In a separate channel in the other direction, veins carry sugars manufactured in the leaf tissue to distribute to the rest of the plant. Veins, like the mountain paths of Incan runners who relayed critical information, are superhighways that are conduits for the essential elements of life. The tiniest veins, barely visible to the human eye, are the workhorses, their narrow passageways reach the outer edges of the leaf and carry critical water and sugars.

Nature displays an incredible array of arrangements for leaf veins. Grasses have parallel veins, palms have radiating veins, and leaves of other plants have branching veins. Plants have been on land for more than four hundred million years, perhaps seven hundred million, certainly hundreds of millions of years longer than mammals and much, much longer than humans. Plenty of time has elapsed for evolution to fine-tune vein arrangements that are the most resilient and least fragile at the lowest cost.

The ginkgo tree, with its fan-shaped leaves and foul-smelling fruit, is the oldest living tree species known on Earth. It lived alongside the dinosaurs and today lines city streets from Seoul to New York. The ginkgo came early in evolution's experiments with veins in leaves. Its distinctive veins make recovery from damage a difficult proposition. The veins radiate from the base of the leaf, in some places splitting into two. But there are no smaller veins that connect across the radiating ones, as if a city had only avenues with no cross streets. If a break or an insect severs one of the veins, the entire part of the leaf from that point onward loses its source of water. The vein architecture is open to attack. To make up for this vulnerability, the ginkgo's leaves contain acids to ward off pests. Early ferns, another ancient plant lineage, share the same primitive vein architecture as the ginkgo tree.

Options for leaf vein designs exploded following the greatest extinction on Earth about 250 million years ago. A huge diversity of flowering plants, with a vast variety of vein patterns, came on the scene after the demise of most other species. Veins in the leaves of oak and birch trees shoot off a central main vein in a fishbone pattern. In a maple tree, the main veins radiate from a central point. But a common feature of the leaves of flowering plants, unlike the ginkgo and early ferns, is a network of smaller veins that connect across the thicker, primary ones. There must be some evolutionary advantage to the non–small world strategy, even though it costs the plant in energy and materials to grow the extra veins.

High-resolution images of microscopic veins provide detailed pictures of vein structures with insights into what that advantage might be. One of the researchers using these tools is Eleni Katifori, a physicist who devotes her career to unraveling the secrets of survival in complex biological networks. One question she has pondered is why so many networks in nature have links that form loops, essentially creating multiple ways to connect nodes, rather than a seemingly more efficient treelike structure that doesn't waste energy and materials on redundant links. Insect wings, coral skeletons, blood-carrying veins in the retina, and leaf veins of flowering plants are some examples. So too are some human-constructed networks, notably city plans for streets, that do not fit Barabási's requirements for small-world status.

Judging from the holes in the leaves of basil and kale in my garden, one of a plant's key survival strategies is to keep water and sugars flowing when under attack from a pest. If the network of leaf veins were like that of a ginkgo tree, a bite out of vein would cut off the two-way flow to all points above and below the attack. A leaf needs a way to reroute the flow, preferably at minimal cost. Katifori concluded that the best solution for a leaf is "hierarchical recursively nested loops," veins that create loops within loops. Nested loops provide not only multiple pathways for water and sugars to travel through veins and connect any two points in the leaf vein network. They also provide multiple pathways to travel to each of the points along each of the alternative paths. In other words, a nested loop provides a lot of options to get from point A to point B. If damage occurs to a vein, the flows can reroute and continue to spread out across the entire leaf. The network becomes more like a web or mesh than a branching tree.

High-resolution images of microscopic leaves confirm a pattern of nested loops in real leaves, as does the yellow fluorescent dye that Katifori and colleagues inserted in the leaves of a lemon tree and a ginkgo tree. In the ginkgo leaf, with an intentional rip in its main vein, the dye does not spread beyond the rip. In the lemon leaf, in a matter of minutes, the flow goes around the rip and the damage hardly impedes the flow. The dye continues to spread throughout the leaf. The loops made the lemon leaf more robust to the tear than the ginkgo leaf. Another advantage of the strategy—veins can accommodate different volumes of flow that vary with the sunshine, rainfall, or temperature. Nested loops just reroute the flow if there is too much for the veins to handle.

Delicate dragonfly wings also have nested looping veins. Their challenge is to keep strength while minimizing weight even if wind or some other calamity tears the wings. The loops create

Gingko leaf (left) and leaf with loopy vein network (right)

webs that act as cross braces, like the braces that hold up the Eiffel Tower or the George Washington Bridge. If some veins get broken, the other braces will be there to keep the wing strong.

The prevalence of nested loops in nature makes the point. Evolution favors redundancy in an uncertain world of insect bites, fluctuations in rainfall, and wind. Nested loops of fine veins make leaves and dragonfly wings resilient to both random and targeted attacks, unlike the hub-and-spoke, small-world networks of sexual partners and the World Wide Web.

But redundancy comes with costs, and nothing is more miserly than evolution. If the cost turns out to be more than the payoff, evolution will find a different strategy. To build additional veins in the interest of redundancy, the plant needs to allocate materials and energy. The strategy is to make redundant veins to create a mesh with narrow, small ones that don't cost much for the plant to build. Evolution has favored looped networks at a small cost to reconcile the small-world trade-off between efficiency and protection against danger in an uncertain world.

MODERN NETWORKS LEARN TO BE LOOPY

AT&T officials did not appreciate the brilliance of a leaf's strategy at the height of the Cold War. They showed the young engineer Paul Baran the door when he suggested a redundant architecture for the military's communication network. The task was urgent. A nuclear confrontation seemed imminent. Baran's job was to advise the U.S. Air Force on a reliable communications network and a way for computer terminals to talk to each other. Communications needed to stay intact even if a nuclear strike took out the military's command and control center.

Baran used a minicomputer to test possible configurations for a communications network: a centralized, single-hub network, which was the norm at the time; a decentralized multi-hub-and-spoke network; and a distributed, weblike network that looked much like the loopy architecture of the leaves of flowering plants. No surprise that the centralized network was the most susceptible to a targeted nuclear strike on the single hub. Also not surprising is that the looped distributed network was less fragile than the small-world, decentralized, hub-and-spoke design. Unlike the other two options, in the distributed network, if a strike hit some nodes, messages could still flow through many different paths. It required more transmission lines, but it would withstand destruction and maintain end-to-end communication under assault. He called his scheme "hot-potato routing."

Baran later recounted the reception when he presented the idea to "a group of old graybeards" at AT&T headquarters. One graybeard stopped Baran mid-presentation. "His eyeballs roll and

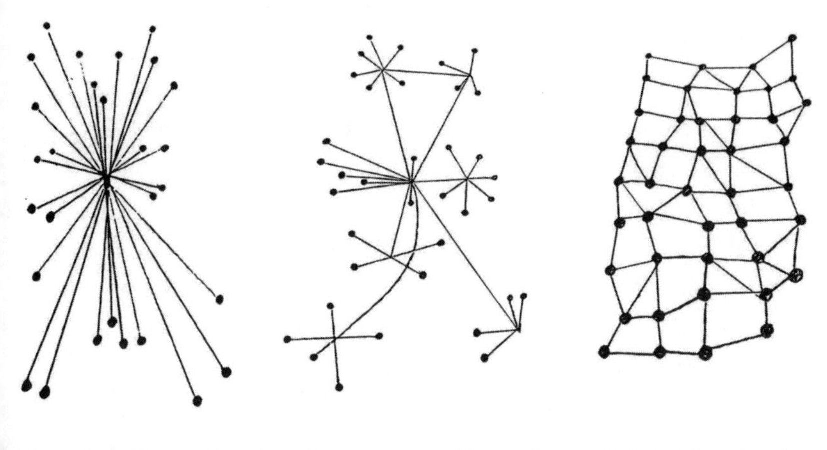

he looks at his associates and shakes his head," Baran recalled. Clearly, the lessons from leaf veins in ginkgos and flowering plants were not on the minds of those who were after the most cost-efficient way to transmit messages. The military and the telephone company ignored his ideas.

For the next decade, Baran pursued his rejected idea. He conducted experiments and published papers and reports on his plans for a "Distributed Adaptive Message Block Network," later termed "packet switching" by another visionary in Britain who independently came up with the same idea a few years after Baran. The goal: to transmit a packet of digital data safely through a network of communication stations so an enemy attack or merely unreliable components couldn't block the flow. Like commuters who find alternate routes to circumvent a traffic jam, Baran's radical idea provided multiple routes for individual packets of information to pass from station to station in order to avoid data logjams, change routes along the way to the final destination, and follow an optimal path.

Baran's reports and papers gathered dust on a shelf, until a graduate student at the University of California in Los Angeles, working with an engineer in the U.S. Department of Defense, saw the value of Baran's foresight. In October 1972, at the Hilton Hotel in Washington, D.C., these future parents of the internet demonstrated the strategy at a conference open to the public. They staged thirty-six computers on a raised floor in a ballroom. People could climb ramps to reach the platform, sit down at a terminal, and connect to computers around the country through telephone lines. As one internet historian wrote, "To most, a computer was a single machine safely stored in an air-conditioned, secured room. The idea of accessing computers around the country from a conference room in a Hilton Hotel was a capability simply unheard of." The demo relied on "hot-potato" packet switching to transmit information between computers. The alternative—dedicated, inflexible routes for each transfer of information—would have choked the network. The internet was born. One of its parents, Vint Cerf, later recalled that he "can't claim any responsibility for having suggested to use packet switching" which came from "other people working independently who suggested that idea simultaneously in the 1960s." Although Baran probably had no inkling of leaf veins and dragonfly wings, the wisdom of a redundant, loopy network for digital communications won out over the skepticism of the early graybeards. The insight ushered in the world-changing age of the internet.

Baran explained why he devoted a decade to his work on resilient networks. "We will soon be living in an era in which we cannot guarantee survivability of any single point," he wrote in 1964. The point applies as much to the interconnected global food trade in the twenty-first century as it did to the Cold War fear of an enemy attack.

NON-LOOPY NETWORKS MEET FOOD RIOTS

When prices of food staples soar beyond the reach of the ordinary person, people get hungry and angry. And when hunger and anger mix with a sense of powerlessness and indifferent rulers, the combination ignites into food riots. Victor Hugo touched on the sentiment in his 1862 masterpiece *Les Misérables* when police drag away the hungry young protagonist, dressed in rags, who in desperation stole a loaf of bread for his sister and her seven children. Food riots recur repeatedly through history, when Parisian women protesting the scarcity and high price of bread ransacked the city armory for weapons and marched to Versailles in October 1789; when grain prices doubled and tripled in eighteenth-century industrializing England as people were pouring into cities from the countryside to labor in factories; during the Irish Famine of the mid 1840s; in New York City, Philadelphia, Boston, and Chicago when food prices soared from a combination of poor harvests and exports diverted to World War I–ravaged Europe; and five decades later when China's Communist Party seized grain from the countryside to feed cities and to trade for imported machines in a "Mao-made catastrophe" that starved millions.

By the mid-1980s, after McLean's containerization had proven its cost-saving efficiency in moving goods around the world, countries were hotly debating trade policies with profound repercussions for the kinds and prices of food available to people all around the world. The U.S. secretary of agriculture at the time, John Block, made his views clear: "The push by some developing nations to become more self-sufficient in food production may be reminiscent of a by-gone era. Those countries could save money by importing more of their food from the U.S. Modern trade practices may mean that the world's major food producing nations . . . are the best source of food for some developing countries." The thinly veiled maneuver to gain customers reverberated in Punta Del Este, Uruguay in 1986, when the United States participated in international trade talks and pushed to remove trade barriers for grains and other agricultural products. The U.S. proposal argued that "due to the greatly diversified sources of agricultural products and the worldwide integration of agricultural trade, it is highly improbable that food shortages . . . would have a lasting or harmful global impact. . . . Good crops in some locations offset the quantitative effects of poor crops in others."

Block's pronouncement reflected the sentiment of agricultural-exporting countries from around the world. Australia had spearheaded the Cairns Group, formed at a meeting in Cairns, Australia prior to the Uruguay meeting. The group aimed to keep agriculture on the agenda in the international trade talks and to lower barriers to exports from their countries. Trade ministers from Argentina, Brazil, Canada, Chile, Colombia, Fiji, Hungary, Indonesia, Malaysia, New Zealand, the Philippines, Thailand, and Uruguay joined Australia in the coalition.

Mexico essentially followed the path that Block and the Cairns Group promoted. "We cannot isolate ourselves in an increasingly interdependent world," Mexican president Miguel de la Madrid told the country in a landmark decision to join the international trade negotiations. The historic change, which the country's largest newspaper called "the most important turn in Mexico's foreign trade policy in almost half a century," ran counter to the principle of self-sufficiency that Mexico had followed since World War II. The subsequent Mexican president continued on the same path, entering into trade-liberalizing agreements with its North American neighbors. The North American Free Trade Agreement went into effect on January 1, 1994, signed into law by U.S. president Bill Clinton.

With trade barriers lowered, cheap corn from the United States flooded into Mexico. The country became a net importer of its staple, an irony considering its place as Vavilov's center of origin for the crop. Yellow corn, used to feed cattle, pigs, and poultry and to make high fructose corn syrup, came from the United States. Mexican farmers grew white corn, preferred for tortillas for human consumption.

In 2005, the underpinnings of the "tortilla crisis" began when U.S. president George Bush promoted independence from foreign energy sources and mandated that a portion of gasoline sold in the U.S. come from renewable sources. To compound the problem, Hurricane Katrina ravaged New Orleans in the fall of 2006 and disrupted oil production in the Gulf of Mexico. Prices for U.S. corn escalated as corn-based ethanol became the solution to meet the mandates and make up for the oil crunch. In Mexico, with higher prices for imported U.S. yellow corn, the livestock industry purchased Mexican-produced white corn to make up the difference. The diversion of white corn caused the price of tortillas to skyrocket. As people could scarcely afford their daily staple, the president had a political crisis on his hands. "Thousands in Mexico City Protest Rising Food Prices" read a headline on February 1, 2007.

Beyond Mexico, prices for edible corn climbed throughout the world. Aided by droughts in wheat-producing Australia and Ukraine and costs for transport from high oil prices, prices of different types of grains also soared. Underlying, longer-term trends of more demand from better-off and urban consumers in Asia and Africa reinforced the price hikes. The result of the global spike in food prices was food riots in 2007 and 2008 throughout the Global South. In Haiti, reliant on "Miami rice" from the United States, demonstrators stormed the palace in Port-au-Prince. In the Philippines, troops in Manila armed with M-16 rifles watched over the sale of subsidized rice. Egypt, Indonesia,

Cote d'Ivoire, Morocco, and Bangladesh were among the thirty-four countries where protests and deadly riots broke out in response to soaring food prices. Block's view of seamless food security for countries relying on imported food started to look like make-believe wishful thinking.

Containers and trade policies, along with mechanization and technological advances on farms, had led to a precipitous decline in the price of food staples after World War II. People could afford to leave farming and purchase their food. Millions of people escaped the poverty trap of low-yield, subsistence agriculture and moved to towns and cities. They made decisions about their jobs and families based on expectations to which they had become accustomed. Then the price spikes revealed the danger of overreliance on cheap imported food when it is no longer cheap.

When prices spiked in 2008 and rioters took to the streets in Mexico, Haiti, the Philippines, and other places around the world, circuit breakers triggered. But the circuit breakers did more harm than good. Governments in countries that export rice and wheat got jittery with the high prices. High prices meant that those producing the rice and wheat would export more of their harvest and there might not be enough for their own citizens. Rice exporters Bangladesh, Brazil, Cambodia, China, Egypt, and India and wheat exporters Argentina, India, and Kazakhstan slapped tariffs on exports. Governments in countries that import rice and wheat also became nervous about the high prices. They took away tariffs on imports to make the products cheaper for their citizens. The net result was counterproductive. Hoarding by exporting countries restricted the global supply. Reduced costs to import increased the global demand. Less supply and higher demand only escalated the prices further in an upward spiral. The circuit breakers that individual countries put in place worked against a collective global good.

Bill Clinton, in a rare moment of humility for a U.S. president, admitted in 2008 that, "We all blew it, including me." Vital food for the world's poor is not a commodity "like color TVs." "It is crazy for us to think we can develop countries around the world without increasing their ability to feed themselves," he told the audience, in sharp contrast to Secretary of Agriculture John Block's advice twenty years earlier.

When prices spiked again a few years later, despite investments and aid for agriculture in response to the 2008 crisis, the feedback reverberated into world-changing geopolitics. In 2010, droughts took hold in wheat-producing Russia, Ukraine, and Kazakhstan. The poor wheat harvest along with high oil prices, increasing demand, and land devoted to growing biofuel converged in another perfect storm. Soaring prices hit hard in Egypt, the largest wheat-importing country in the world. High costs for daily bread added to frustrations from poor governance, water scarcities, and lack of jobs. The combination sparked the Arab Spring, starting in 2010 and lasting into 2011. Uprisings spread throughout the Middle East and North Africa, toppling governments and resetting the world stage.

To date, civilization's experience with an intertwined global food trade is in its infancy. Countries have responded by shoring up their stockpiles and international bodies have debated proposals for trade rules and designated reserves in key places. Collaboration and agreement among so many countries on such a crucial concern follow a long, rocky path.

The cracks in the global food trade reveal the classic efficiency versus resiliency paradox of small-world networks. Containerization and trade have shrunk the global food trade to a small world. Like the World Wide Web and other human-constructed networks, the modern world's global food system has a few hubs that produce essential staples with many spokes to countries around the world. Seven countries—the United States, Canada, France, Australia, Russia, Germany, and Ukraine—export three-quarters of all internationally traded wheat distributed across 170 countries. Export of rice is even more concentrated in hubs, with four countries—India, Thailand, Viet Nam, and Pakistan—exporting three-quarters of the total purchased by 137 countries. Similarly for corn, five countries—Brazil, the United States, Argentina, Ukraine, and France—export three-quarters of the total traded to 115 countries. These hubs of the hub-and-spoke, small-world network form the Achilles' heel of the global food system.

The global food trade brings security in case a calamity strikes within a particular country. Thanks to trade, famines that plagued people throughout history are far less frequent today and more likely to be triggered by political conflict than droughts or disease that strike a locally produced crop. But the small world of the global food trade also brings a different kind of vulnerability and fragility unlike past, more rural, and more disconnected times. In particular, in the Global South, the urban poor have no option but to spend a very high proportion of their income on food. In countries that rely heavily on imported food, they are at the mercy of fluctuations in the international market and faraway catastrophes that cascade around the world. Their frustration and loss of dignity when the price of bread soars beyond their reach, a situation completely outside their control, can trigger violence and topple regimes.

In the modern world, as more people live in cities and purchase their food, dependence on affordable food from faraway sources grows. These city dwellers rely on ships carrying wheat, rice, and other staples that pass through narrow chokepoints. Blockages to the Panama Canal, Suez Canal, or the Strait of Malacca could send prices soaring. On top of those complexities, climate change ramps up the volatility of the food supply with more frequent droughts, floods, and heat waves.

Leaf veins and dragonfly wings suggest possibilities for national governments to consider. The global food trade would be less fragile and less prone to food riots with more redundancy, somewhere between a small world and isolated clusters. At one extreme, complete self-sufficiency for food production would likely result in higher costs; greater vulnerability to droughts, floods, and pests that occur within a country; and less diverse diets, as people could only consume what could be locally produced. At the other extreme, a small world with overreliance on a few countries as the source for food staples is equally vulnerable but in a different way. Price hikes that cascade through the international market can devastate a dependent country and incite violence. The fragility of the modern world's global food trade, as one would expect from a small world, has started to become clear. Nature's voice of experience—a diversity of trade partners, reliance on a diversity of staples, and balance between self-sufficiency and trade even at the cost of efficiency—can be the difference between frustrated, hungry, urban masses banging pots and burning buildings and more stable lives for billions of people around the world.

WHEN DANGER SPREADS ACROSS NETWORKS

The internet, the global food trade, and leaf veins have a common goal: to keep information, food, or water and sugars flowing through their networks without disruption. But when networks act as conduits for harmful diseases, or for falsehoods and dangerous ideas, the goal flips. Rapid action to stem the flow becomes crucial for public health officials and those who work to keep society safe from dangers that exponentially fan out across social networks.

Termites, honeybees, ants, and other social insects face this problem in their networks. If one member carries a potentially lethal virus or fungus into the nest, the insects need to cut off its spread to nestmates as quickly as possible.

Crowded, group living in a colony has its benefits. A group can guard against predators and share responsibilities to find food, get rid of waste, and raise the young. But the strategy has a distinct downside. A disease can spread like wildfire through the social network and destroy the entire colony, particularly if the disease infects the queen.

Curiously, social insects do not often succumb to epidemics, even when they live in close quarters with millions of nestmates. They have a few ploys to compensate for the danger from living in crowded groups. Garden ants secrete a disinfectant and spread it on their nests. Wood ants and honeybees collect resin from plants to sanitize their nests. Termites spread their own feces in their nests to benefit from its antimicrobial properties. When a fungus or disease-carrying mite attacks, healthy members pick spores and mites off their sick nestmates. Honeybees flap their wings in synchrony to create heat that kills the pathogen; dying members leave the nest of their own volition; healthy insects drag corpses away from the nest. Somehow, the nestmates know to take on these behaviors in the interest of the entire colony.

Most remarkably, these tiny creatures do more than change their individual behavior to stem the spread of disease through the colony. They also collectively manage the architecture of their social networks to curtail the spread. Their networks are far from random. Each nestmate does not have an equal chance of sharing food or communicating with any other nestmate. Rather, the networks are highly compartmentalized into clusters. Older individuals live on the periphery of the colony and carry out risky tasks of foraging for food and guarding against predators. Younger individuals live more safely on the inside and care for the brood and the queen at the center. Those who carry out wastes live together at a distance from other groups. Their social networks are highly modular. Those with similar roles interact mostly with each other, like cliques in a high school lunchroom.

If a pathogen somehow gets through the defenses, the insects rapidly signal to each other and adjust their social network. Infected individuals become socially isolated and don't interact with healthy nestmates. Those who acquired immunity from low-dose exposure to the pathogen as they cared for the sick form a protective wall around the queen. The modular network enables

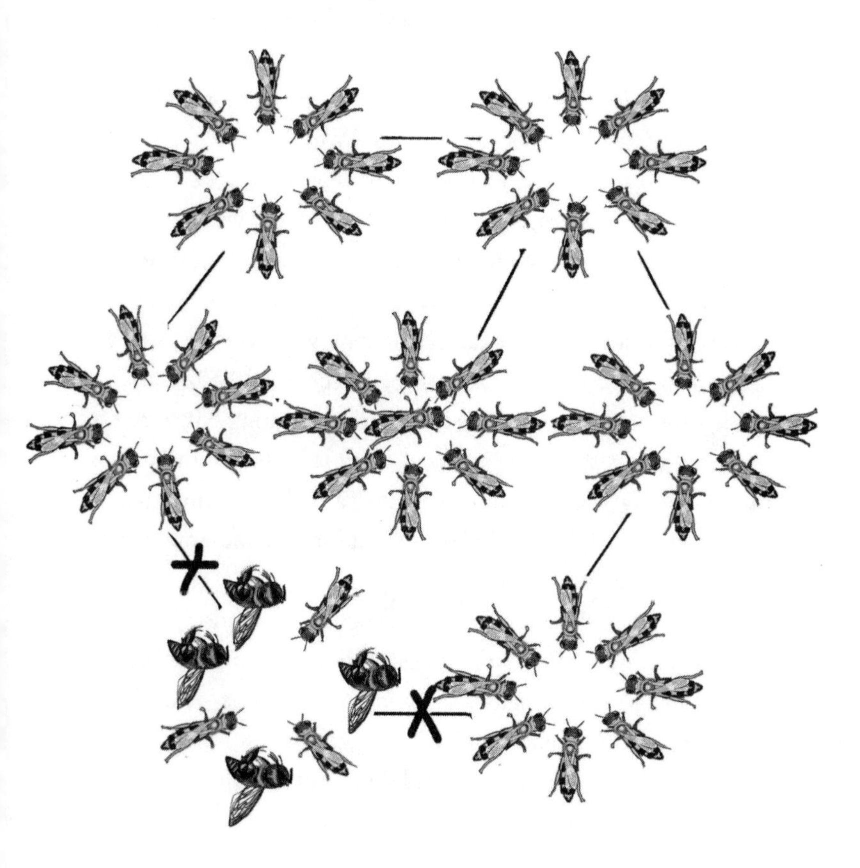

clusters of unexposed insects at the center to close their entrances and move farther from the other clusters. Instinctively they know that the architecture of their social networks can be the difference between warding off a deadly disease and the death of the colony.

Since people began to live in crowded conditions, human civilization acquired the same problem as honeybees, ants, and termites. Epidemics, which small, mobile groups of hunters and gatherers could not sustain, became part of our species' existence and continue until today. In the stew of early agrarian life, as people crowded into settlements and lived beside livestock and rodents, viruses and bacteria mutated and jumped from animal to human. Smallpox was one of many diseases that were born as civilization took root—among others were malaria, tuberculosis, and measles.

Thousands of years passed before smallpox no longer plagued humanity. It remained until the World Health Assembly declared in 1980: "The world and its people have won freedom from smallpox . . . a most devastating disease sweeping in epidemic form through many countries since the earliest time, leaving death, blindness and disfigurement in its wake." The heroic effort to eradicate the deadly smallpox virus from the face of the earth involved grit, leadership, centuries of discoveries, and hundreds of thousands of health workers fanning out to remote villages across the world to track every last case. In the process, these courageous men and women learned over time that the networking strategies of honeybees, termites, and ants work as well for humans as for insects.

The story of the end of smallpox begins thousands of years ago. The details are murky, but the deadly virus likely evolved between three and four thousand years ago from cowpox or camelpox. Between its emergence and its ultimate demise, smallpox epidemics roiled civilizations around the world and killed many millions. The virus took root in the human population around 500 BC in the Ganges River valley. Throughout the next two millennia, epidemics rolled in almost like clockwork every five or ten years as new generations with no immunity became susceptible to the virus in Europe, Asia, and Africa. Smallpox epidemics decimated indigenous populations in the Americas who had never before been exposed to the disease until Spanish and Portuguese conquistadors arrived in the early sixteenth century. In India, devout Hindus prayed to the deity Sitala, the "cool one," to protect them from the disease that studded the body with burning pustules, decayed flesh, killed half of its victims, and disfigured and blinded those lucky enough to survive.

Like insects who acquire immunity from low-dose exposure to infected nestmates, folk wisdom over the centuries figured out ways to boost immunity and reduce chances of contracting the disease. The Chinese inhaled powder ground from pustule crusts shed by recovering patients. Africans rubbed pus from a smallpox lesion into a cut in the skin. In Europe, parents tried to protect their children by sending them to collect crusts of lesions from smallpox victims. But the practices were not foolproof. Some people died from mild exposure to the virus and inoculated patients could spread the virus to trigger an epidemic.

A breakthrough came from an eighteenth-century milkmaid in the English countryside who repeated common lore to a rural country doctor, Edward Jenner. She was sure that she would never get smallpox since she had previously recovered from cowpox, a relatively benign disease that causes mild skin rashes. Milkmaids had figured out that if they survived cowpox, they would be spared the "speckled monster," as smallpox was known in eighteenth-century England. The hint that cowpox could make people immune to smallpox sparked Jenner's interest. With methodical and spectacular experiments to inject pus from cowpox lesions into the arms of twenty-three people, including his own children, Jenner proved the case. He endured much skepticism and ridicule, but eventually inoculation with relatively benign, live cowpox pus became the norm in Europe and the Americas. Parents no longer needed to worry about the dangers of inoculation with low doses of the live smallpox virus. The nonlethal cowpox virus worked just as well.

People came up with ingenious ways to obtain live cowpox pus for inoculations from lesions. One solution to keep the virus alive through ocean journeys: round up orphans to form an arm-to-arm living chain and keep the virus alive through the journey. Infect one orphan with cowpox, transfer to another when the lesions in the first start to scab, and so on until the human carriers deliver the virus at the other shore. Another solution: lead a calf from door to door and scrape a bit of pus from its flank for each inoculation.

A century after Jenner's experiments, Pasteur—the French scientist famous for discovering that microscopic germs, not the weather or position of the planets, cause disease—stumbled upon a solution. Pasteur's assistant accidentally left a live culture of cholera bacteria on a bench while Pasteur was enjoying his summer vacation. The bacteria died. When Pasteur returned, he injected chickens with the then-dead bacteria on a whim to see if it would make the chickens immune to cholera, which was ravaging Europe's chicken population at the time. Lo and behold, it worked. Injected bacteria and viruses did not need to be alive to bolster immunity. Vaccines, which Pasteur named after Jenner's Latin name for cowpox, *variolae vaccinae*, could be manufactured in a laboratory from germs weakened by high temperatures or exposure to air. Orphans and calves' flanks were no longer required. The end to civilization's millennia-long struggle with the deadly disease was in sight.

By 1950, vaccination campaigns had eliminated smallpox from North America, western Europe, the Philippines, and some countries in Central America, excepting cases imported by travelers from places where the disease still occurred. The disease was still circulating in most of the developing world, causing deaths and danger that the disease could again spread. In 1959, the World Health Assembly, composed of representatives from countries around the world, embarked on mass vaccination programs in places where smallpox was most prevalent. A technique to produce a freeze-dried smallpox vaccine that was stable in hot climates aided the ambitious endeavor. Even with these efforts, smallpox was still rampant. The World Health Assembly doubled down with an intensified eradication program in 1967 and called for more mass vaccination campaigns. The target was to vaccinate at least 80 percent of the people in the worst-affected countries.

Donald Henderson, the man who steadfastly steered the eradication effort with determination and humility, later explained that the intensified campaign accidentally stumbled upon the network-altering strategy that ultimately defeated the virus. "There was a delay in the delivery of supplies for the mass-vaccination program in eastern Nigeria, and an energetic U.S. adviser, William H. Foege, organized an interim program: he searched out smallpox cases and vaccinated thoroughly in a limited area surrounding each case. The mass-campaign supplies arrived only a few months later, but by then there was no detectable smallpox in eastern Nigeria. And less than half of the population had ever been vaccinated. This result and similar experiences in other places led to . . . what became known as surveillance-containment: improved search and detection as speedily as possible, isolation of patients and vaccination of every known or suspected contact around them. The procedure sealed off outbreaks from the rest of the population."

With the success of the surveillance-containment method, hundreds of thousands of health workers journeyed over dirt roads to far-flung villages across Brazil, India, Bangladesh,

Nepal, Pakistan, Botswana, Sudan, Nigeria, and other countries around the world to track down each case of smallpox, seclude the patient, and vaccinate every person known to have physically come in contact with the carrier. The new strategy involved diligence, house-to-house searches, payments to local people to lead the health workers to infected persons, and guards to keep sick people from leaving their homes. The efforts paid off.

On October 31, 1977, Ali Maow Maalin, a twenty-three-year-old cook in a town two hours' drive from Mogadishu, Ethiopia, was the last person to catch smallpox from another person, in his case a six-year-old girl or her one-and-a-half-year-old brother while giving the children a fifteen-minute-long ride to a clinic. Maalin was lucky. He recovered from the often-fatal disease and devoted the rest of his life to eliminating another crippling disease, polio, in his country. He died from malaria in 2013.

With the 1980 declaration, the world was free from smallpox, the devastating disease that killed, blinded, and disfigured millions in waves of epidemics throughout history. No more fear of outbreaks, no more vaccination campaigns. The virus was obliterated from the planet.

The surveillance-containment strategy, the key to the success of the eradication program, created a ring of immunity around each infected person. The virus could not spread beyond the social network. Managing the network proved to be a more effective strategy and more efficient deployment of scarce vaccine stocks than mass vaccination of the population. Honeybees, termites, and ants use these same strategies—low-dose exposure to the pathogen to build immunity and isolation of infected nestmates—to curtail epidemics in their crowded colonies. Humans, through accidents and wit, learned the wisdom of these strategies.

Eradication of the smallpox virus was an amazing, difficult, and daring feat. But eradication of other deadly diseases is an even more daunting prospect. The man on the frontlines of the smallpox campaign, Henderson, sounded the alarm about "excessive optimism" in the quest to eradicate other diseases that emerged when humans began to live more like social insects than like small solitary bands of animals. He summed up the reality from the field with the warning that "prospects for eradication appear far more optimistic from the vantage point of a laboratory or an office in a university ivory tower."

Henderson was well aware of the previous attempts to eradicate diseases. The failure rate was 100 percent. First there was hookworm, a blood-feeding parasite that burrows into the walls of the victim's intestine to cause listlessness, anemia, and stunted growth. The philanthropist John D. Rockefeller took up the cause in 1909 through privies, medications, and campaigns for people to wear shoes to avoid picking up the parasite from the soil. The program halted hookworm in the southern United States, but people couldn't comply in the rest of the world, and the program failed badly. Then the Rockefeller Foundation set its sights on ridding the world of yellow fever, a virus carried by mosquitoes that causes chills, fever, aches, jaundice, and often liver damage and death. But the virus proved evasive. It lurked in a wild reservoir—monkeys in the jungle—and there was no way to eradicate the disease. An attempt in the 1950s to eradicate yaws, a disease that causes painful skin lesions, met a similar fate. The problem was that the infection lingers for weeks and spreads after treatment with penicillin. Health workers could not isolate seemingly healthy people for so long. In yet another attempt, four years before the 1959 global declaration to banish smallpox, the World Health Assembly began a program with "its ultimate objective the world-wide eradication of malaria." Despite initial success from spraying the super insect killer DDT, malaria roared back in many tropical countries. Mosquitoes and the malaria-causing parasites that they host evolved to resist DDT. It remains one of the world's most debilitating diseases. In all these well-intentioned but unsuccessful attempts at complete eradication, formidable disease-spreading networks – involving monkeys, mosquitoes, and entrenched human behavior – were just too tough to break.

Perhaps in the near future a small handful of diseases will join a list of banished diseases that have plagued humanity. Rinderpest, a measles-like virus that devastated herds of cattle, buffalo, and other cloven-hoofed animals for millennia, became the first eradicated animal disease in 2011 after more than a decade of vaccination campaigns. Polio is in its end-game as of this writing. Guinea worm, a painful disease caught from larvae in water fleas that burrow into skin, and measles could follow. Malaria, with the power of Bill and Melinda Gates behind an eradication attempt, could follow suit in coming decades. That leaves a long list of others—tuberculosis, influenza, cholera, AIDS to name just a few—whose complex biology doesn't easily cooperate with moonshot hopes for eradication.

To make the problem even thornier, new diseases keep emerging as people and animals mix on farms and in forests as they did when crowded civilizations began. In our connected world, pathogens spread fast and far as people travel. Three hundred and thirty-five diseases emerged from obscurity between 1940 and 2005, among them West Nile virus, which originated in Uganda and appeared in New York City in 1999; Lyme disease, which spread across the United States since the 1980s as burgeoning deer populations and regrowing forests created an opportunity for pathogen-carrying ticks; and Zika, which originated in Africa and appeared in mosquitoes in Brazil in 2015. The number of new diseases making the jump from animals to humans is on the rise.

The race is on, as of this writing, for a vaccine to counter the economy-crippling destruction from the novel coronavirus that killed hundreds of thousands of people in a matter of months from late 2019 into 2020. The window of opportunity to emulate termites and honeybees by isolating coronavirus-infected clusters closed as the virus spread from Wuhan to all continents except Antarctica within a matter of weeks. China, after initially downplaying the urgency, came closest to following that strategy as the government used its authority to lock down cities, keep people from leaving their homes, and isolate infected people and those who came in contact with them. Singapore, Taiwan, South Korea, Vietnam, and Hong Kong put similar measures in place. Their success in reducing coronavirus cases shows the wisdom of cordoning off clusters in modular networks.

Whether the module is a country, a community, or a household—rather than groups of social insects clustering apart from each other in a nest—evolution's experience enforces the wisdom of quick, collective action to stem the spread before the virus infects the next module. During the catastrophic 1918 Spanish flu pandemic, the Board of Health of a small mining and farming town in the mountains of Colorado proved the point with a bold decision. Before the virus hit the town, they put the people of Gunnison on lockdown. No one could come in, children could not go to school, and churches suspended services for four months. The virus spared them. Only a handful of people died from the deadly flu, while scores perished in neighboring towns.

On a global scale, the World Health Organization, the international body designed to coordinate countries to respond to such emergencies, has long known the effectiveness of swift action. Their commonsense advice for rapid containment and quarantines essentially follows the network strategies of social insects as much as possible in human society. In today's interconnected, fast-paced world, when pathogens like the coronavirus can spread around the world in a day, the strategies take on extra urgency.

Vaccines for all the diseases that have emerged, and are likely to emerge in the future as new pathogens continue to jump from animals to people, would require funds and effort beyond imagination, not to mention a good dose of luck to be as effective as the ones for smallpox and polio. Even smallpox has a small but non-zero chance of making a comeback, a horrific prospect now that people no longer get vaccines and have no immunity Complacency or, worse yet, people's irrational resistance to vaccines can turn the clock backward. Despite humanity's spectacular success in eradicating smallpox, epidemics will remain part of the human experience. Only ingenuity, innovation, and humility can keep the problem at bay.

As the serendipitous discovery in the smallpox campaign proved, breaking the flow of harmful viruses and bacteria means building immunity and severing links at critical points in the social network. Where and between whom those critical points exist depend on the structure of the social network. If the social network has clusters like ants, termites, and honeybees, break the link between infected and healthy clusters. If super-spreaders act as hubs like Typhoid Mary, isolate the spreader. If people and pathogens collide more randomly, build a ring of immunity around infected victims as Foege and Henderson realized when smallpox vaccines were scarce. Answers to critical questions depend on the network's structure: who to immunize with what priority; where to deploy scarce supplies of personnel and vaccines; whether to impose draconian measures to isolate infected people from their daily lives. With guidance from colony-dwelling creatures who have lived with the threat of epidemics for millions of years, our species is still learning how to manage our networks when diseases inevitably spread in our interconnected world.

What is life on Earth's third piece of advice for the modern world, based on its experience over millions of years with complex networks? It might go something like this: Construct your human-made networks. They are the foundation of modern civilization. Networks maintain the flow of information, food, electricity, and other necessities of modern life. But be aware that danger also lurks from cascading failures that can ripple through your networks. To keep necessities flowing, resist the tendency toward constructing networks with few hubs and many spokes. These small-world networks will inevitably emerge as networks grow, especially if there are few constraints on how your networks expand. The inherent structure of small-world networks will only make them fragile when, inevitably, disaster strikes a hub. Paul Baran's insights with his distributed plan for the internet and evolution's experience with leaf veins apply to all the networks you rely on for your necessities. Countries beware of sole sourcing the staples that feed your population. Businesses keep options for suppliers to your supply chain. Build redundant pathways into your networks, even at the price of some efficiency.

Networks also carry hazards. Diseases and harmful rumors can flow through social connections and cascade out of control without much warning. As honeybees, ants, and termites instinctively know, and the incredible campaign to eradicate smallpox shows, tweaks in the architecture of a social network can stem an epidemic. Learn strategies from the social insects, who have faced the same problem for millions of years. Act quickly and decisively to cordon off infected parts of the network. Use your scarce resources wisely to target your vaccines to those most likely to be infected. With attention to nature's experience and forward-thinking planning, the uncertain, dragon-filled modern world can benefit from networks but avoid their pitfalls.

5

ONE SIZE FITS NO ONE

Make Decisions from the Bottom Up

WHO owns the fish in the sea? Who has the right to dump pollutants or breathe virus-laden particles into the air? How much water can someone pump from a groundwater aquifer that many people share? No one can put a fence around fish, air, or an aquifer. But when someone hauls in a fish, pollutes the air, or pumps a bucket of water from the ground, less remains for the next person. This is the problem of common resources that are owned by no one and shared by everyone.

The "tragedy of the commons" affects all of humanity and takes on urgency in our inter-connected world where one person's action ripples across the globe. It requires wisdom and collective action to resolve. Individuals of other species also need to join together around common goals, whether to build nests, protect one another from predators, or collectively gather food. How evolution coaxed individuals to take actions that benefit everyone suggests possibilities for our own species.

More than two thousand years ago, Aristotle remarked on a reality of human nature. "For that which is common to the greatest number has the least care bestowed upon it. Everyone thinks chiefly of his own, hardly at all of the common interest; and only when he is concerned as an individual." People make decisions in their own interests, not for the good of the group.

In the second half of the twentieth century, much-needed sanitation, vaccines, improved nutrition, and public health measures around the world saved children from dying and extended people's lives. Fewer deaths meant more mouths to feed and more demand for "that which is common" in the form of water, minerals, waste disposal sites, and other resources. Books, which greatly influenced my thinking as an impressionable young college student, screamed alarming titles. Paul Ehrlich penned *The Population Bomb* and Donella Meadows *The Limits to Growth*.

In 1968, the same year Ehrlich hit the airwaves popularizing a scary neo-Malthusian specter of a world overrun by too many people, a professor of biology in Santa Barbara, California, published an essay outlining how society should manage common resources.

Garett Hardin framed "the population problem" as too many people scrambling for too few common resources. He described the tragedy of the commons through the analogy of a herdsman grazing his cattle on common pasture shared by many herdsmen. "Picture a pasture open to all," he wrote. "It is to be expected that each herdsman will try to keep as many cattle as possible on the commons." His logic flows that "the rational herdsman concludes that the only sensible course for him to pursue is to add another animal to his herd. And another; and another. . . . But this is the conclusion reached by each and every rational herdsman sharing a commons. Therein the tragedy. Each man is locked into a system that compels him to increase his herd without limit—in a world that is limited. Ruin is the destination toward which all men rush, each pursuing his own best interest in a society that believes in the freedom of the commons. Freedom in a commons brings ruin to all." "The tragedy of the commons"—the title of the influential essay—today is everyday parlance to depict the fate of shared resources from office refrigerators to the atmosphere and oceans that serve as dumpsters for unwanted by-products of civilization.

Hardin's essay and later writings alluded to a twofold solution to the tragedy of the commons. He aimed to alter the equation for rational decisions. "The rational man finds that his share of the costs of the wastes he discharges into the commons is less than the cost of purifying his wastes before releasing them," he reasoned. One way to resolve the tragedy abandons the notion of commons through private ownership, much as private property and fences foreclosed common people from gathering food and accessing land in the early days of the Britain's Industrial

Revolution. The cost of trespassing would cause a rational man to stay off the property and private owners would manage their resources sensibly. For other commons that are not amenable to fences and property deeds—oceans, the atmosphere, fisheries, and aquifers as examples—higher authorities need to "allocate the right" to use the resource. Prospects of punishments, fines, and jail time would change the decision-making equation for the rational man who is thinking about abusing the commons. As for the problem of overpopulation, Hardin's solution involved "relinquishing the freedom to breed."

Hardin's ideas became nearly dogma in development circles and resource management agencies and among environmentalists. Textbooks to train future managers devoted entire sections to the tragedy of the commons and claimed that "the parable of the commons tells us that people naturally tend to overconsume nature's bounty." Policy advisers adopted Hardin's prescriptions for privatization and state control over common resources. Governments followed this advice. The 1976 Policy for Canada's Commercial Fisheries, for example, referred directly to the tragedy of the commons and led to government-issued licenses and limits on the number of fishing boats. Hardin's idea profoundly influenced international rangeland experts who advised governments in the Middle East and elsewhere to settle nomadic populations in the name of degradation of the steppe. Legislation throughout the developing world transferred forests, fisheries, pastureland, and other resources to government ownership. By the 1970s, top-down state control or exclusionary privatization seemed the only option to dodge devastating depletion of Earth's resources.

In the meantime, Elinor Ostrom, an unassuming woman from Los Angeles raised in Depression-era poverty, was quietly compiling data and analyzing how people actually manage common resources in the real world. In 2009, she stood on the stage in

Stockholm to accept the Nobel Prize in Economics for her work. Her conclusions upturned Hardin's notion that only top-down directives from centralized authorities can solve the tragedy. Ostrom didn't deny that tragedies in commons occur. She added more tools to the tool kit to deal with the problem. People, she found, can effectively organize themselves to manage their own commons if conditions are right. As birds form flocks and ants carry food to nests using bottom-up principles of communication and collective action, people can solve their own problems.

TRAGEDIES AND COMEDIES OF THE COMMONS

I was sitting at my desk when the news came across that Elinor Ostrom was the 2009 recipient of the Sveriges Riksbank Prize in Economic Sciences in Memory of Alfred Nobel. I jumped up in elation. Lin, as she liked to be called, was the first woman to win the prize. And she was not an economist. She was a political scientist who rode in police cars and tromped through forests to collect data. Most critically, she questioned conventional wisdom through painstaking and careful work about how people actually behave in the real world. She did not look favorably on conclusions from abstract and theoretical notions formulated behind a desk. Lin was deeply loved by her students and colleagues not just for her intellect. She was generous in spirit, completely unpretentious, and lived a life full of meaning. "You go girl" was my fist-pumping, instinctive reaction to the news.

Before Hardin's tragedy of the commons caught the attention of governments and the public, Ostrom cut her teeth on a potential tragedy in the making. The city of Los Angeles was in trouble. During the first half of the twentieth century, as population grew and the city expanded, the level of the groundwater table had dangerously declined, land subsided, and seawater intruded on the coast. Hundreds of water users were pumping from the basin, from individual farmers to municipal water utilities. The below-ground boundaries of the groundwater basin did not match any political jurisdiction. Water was a common resource and no one had authority to restrict use.

Ostrom documented over several years the morass of water-managing institutions that clumsily but assuredly collaborated toward a resolution in which they all played a role. They agreed to use less groundwater and report how much they used, replenish groundwater through injecting freshwater along the coast, and adjust water allocations over time. The agreements arose from hundreds of face-to-face meetings and conversations, with plenty of conflicts along the way but without a central authority imposing rules and restrictions. Ostrom later wrote: "I learned early . . . that individuals facing such problems do not always need an external authority to extract them from their tragedy. When they have arenas in which they can engage with one another, can learn to trust one another, can draw on sources of reliable data, can ensure monitoring of their decisions, can create new instrumentalities, and can adapt over time, they are frequently, though by no means always, able to extract themselves from these challenging dilemmas."

Ostrom's next major undertaking brought her head-to-head with conventional wisdom of the day. In the 1950s, advocates of metropolitan reform for American cities claimed that too many small policing units led to fragmentation, chaos, and ineffective service. One big consolidated police force would be more efficient and more effective. Ostrom, her husband and career-long collaborator Vincent Ostrom, and students took on the task to test the idea. They rode in police cars, observed officers, and talked to thousands of residents in Indianapolis, Chicago, St. Louis, Grand Rapids, and Nashville. Their conclusions ran counter to conventional wisdom. Smaller police units provide better services at less cost, with fewer crime victims, victims who are more likely to call the police, and citizens who give higher marks to their police forces. "In this set of studies, no one found a *single* case where a large centralized police department was consistently able to outperform smaller departments serving similar neighborhoods," was the conclusion. Police officers and residents can carry out their tasks with ears and eyes close to the ground more effectively than a large, centralized authority. Well-intentioned reformers had based their advocacy on untested theories.

Ostrom might have had the untested theory of the tragedy of the commons on her mind when Hardin came for dinner at her and Vincent's home in Bloomington, Indiana, during a visit in 1976. Hardin's case for the tragedy sounded compelling, but he hadn't done much homework with actual evidence. His metaphor of herdsmen grazing cattle on pastures didn't ring true. In feudal England and in many places around the world today, landowners and communities set complicated rules for herdsmen about how many of which types of animals can graze in what season. No doubt common resources that are open to everyone without any oversight, as the ocean and air have been for millennia, invite abuse from free riders who shirk the responsibility of cleaning up their messes. But perhaps even these problems have more solutions than top-down private ownership or rules set by a central authority.

Hardin's prescriptions didn't square with Ostrom's on-the-ground observations of water management in Los Angeles. She had witnessed that sometimes people can come together from the bottom up to set and enforce their own rules. And she had seen from her studies of police forces in American cities that widely accepted, purportedly conventional wisdom that looks sensible on paper might not be correct when tested against the real world. She viewed Hardin's popular ideas as overly restrictive simplifications of the many ways people organize themselves to solve problems. And alarms went off about Hardin's top-down, draconian solution to the prospect of too many people polluting and gobbling up the commons. Hardin argued that every man and woman should be sterilized after one child. Following the dinner, both Lin and Vincent expressed to a colleague a "deep concern about Hardin's 'totalitarian' birth-control policy."

After decades of studying fishers, farmers who irrigate their fields, and people who depend on forests for their livelihoods, Ostrom and her many colleagues formulated an alternative to Hardin's tragedy of the commons. The "drama of the commons" sometimes ends in tragedy and sometimes in comedy. In some cases people self-organize to effectively manage their common resources over long periods of time. In other cases, Hardin's predicted "ruin" is reality.

A tiny village called Torbël, perched on a mountainside in the Swiss Alps, gave Ostrom one glimpse into how people self-organize to guard against tragedy. When peasants settled in Torbël in medieval times, they collected wood from the forest to build houses and keep warm through the cold, harsh winter. They collectively grazed their animals on alpine pastures in the warmer months and fed their animals with hay harvested from their fields in the snowy winter months. On February 1, 1482, twenty-two Torbël residents gathered to draw up a voluntary agreement to specify how they would share the common resources. The rules, recorded in Latin on parchment, forbade villagers from granting rights to outsiders to use the communal alp for grazing animals or to fell timber. No resident could send more cows to graze on the alp than he could feed during the winter with hay harvested from his own meadows, a rule with severe fines that the village enforces to this day. Through time, the villagers codified more rules about the use of horse trails and cow paths. Without outside authorities, recorded in meticulous documents over centuries, the community devised, enforced, and refined rules to keep their common pastures safe from overgrazing and forests free from unsustainable felling.

The strategies that coastal communities employ to guard against depleting their fish stocks gave Ostrom another glimpse into alternatives to tragedies. In the small Turkish town of Alanya on the Mediterranean coast, fishermen can voluntarily join a cooperative. They agree on rules each September at the beginning of the migratory fish season. Each fisherman agrees to his fishing locations and deposits a contract with the local officials. Although the contracts have no legal status, the community does not look favorably on those who violate the terms. They solve disputes in the coffeehouse with social snubs and sometimes with threats of violence.

Perhaps Ostrom gained her greatest insights from the lush green rice paddies that blanket landscapes throughout Asia. To get water to the paddies, people engineered complex irrigation systems over centuries with dams, tunnels, and water-diverting structures. The systems can only function with rules for allocating water among users and collectively sharing labor, materials, and costs to build and maintain the systems. In Nepal, governed for centuries by princes, farmers devised and managed tens of thousands of irrigation systems without any central authority. Each community developed its own rules to ensure that everyone contributes to the effort and to punish free riders who take more than their share of water. In the mid-1950s, the government of Nepal established a department of irrigation with five-year plans to construct large, centrally managed systems funded by international development banks and donors. Comparison of these two strategies revealed that a higher proportion of the farmer-managed systems remain in good physical condition, with more water for farmers and higher yields, than the centrally managed systems. It stands to reason. Farmers participated in crafting rules to suit their systems and invested in the success of those systems, unlike the systems managed in a faraway capital over which they had little control.

These and thousands of other local case studies led Ostrom to ask a fundamental question. Tragedies of commons occur all around us, with corrupt officials grabbing resources, overexploited fisheries, sinking groundwater levels, and polluted air and water. But in some cases, people manage to share the commons and prevent Hardin's "ruin" with mutually agreed rules that endure over the long term. Why do some dramas end in tragedy and some in comedy?

Ostrom settled on eight "design principles" to capture the key differences. Among the eight: rules to clearly define who has the right to use the resource, as the Torbël residents mandated; a credible way to monitor the condition of the resource, as the Nepali farmers tracked how much water each farmer withdrew from the canal; mechanisms to resolve conflicts, as the fishermen practiced in the Alanya coffeehouse; and rights for the community to set their own rules, based on the ecology and culture of the place and people, without top-down imposition from officials.

No single set of rules determines whether the drama of the commons ends in tragedy or comedy. As Ostrom explains from decades of deciphering how communities use their resources, "We have not found a particular set of collective-choice rules to be uniformly superior to others. We and other scholars have consistently found, however, that rules developed with considerable input of the resource users themselves tend to achieve a higher performance rate than systems where the rules are entirely determined by external authorities." She argued that no one can draw a fixed-in-time blueprint or devise a one-size-fits-all set of rules that can work in all places at all times. As the Inuvialuit know from cracks in the ice, local people are the best judges. They can see, feel, and smell their surroundings better than a far-off authority.

Ostrom's bottom-up challenge to Hardin's top-down paradigm upturned how governments, donors, and development banks manage forests, fisheries, and other common resources that we all depend upon for our survival. Hardin's idea of centralized management had not stemmed the assault on common resources. Toward the end of the twentieth century, fish stocks were on a sharp downward trend, forests were disappearing, and groundwater levels were falling in many places around the world. The buzzword for the remedy was "decentralization." Local authorities, rather than central government, should be responsible for making and enforcing rules about common resources.

By the beginning of the twenty-first century, many central governments got on the decentralization bandwagon and delegated power to lower-level authorities. Thousands of local councils for fishers to "take over their fishery, own their boats, run their businesses, negotiate prices and working conditions," in the words of a Canadian official, sprang up along the coastlines of Canada, the United States, Norway, the Philippines, the Netherlands, Japan, Bangladesh, and other countries. In the mountain kingdom of Nepal, a previous 1957 law had nationalized all forests and usurped local management systems. People viewed the trees as state property and cut them down without restraint. By the 1990s, new legislation placed forestry management back in the hands of local communities and trees rebounded. Today, Nepal is a leader in decentralized forest management. Indonesia, India, Bolivia, Nicaragua, Uganda, and other countries followed suit with varying degrees of success. Decisions about groundwater, a common resource particularly susceptible to tragedies, also fell under the decentralization banner. Representatives of one hundred countries convened in Dublin in 1992 and adopted the principle that "water development and management should be based on a participatory approach, involving users, planners, and policymakers at all levels."

Ostrom's bottom-up tool became part of the tool kit for national governments and international agencies to manage potential tragedies of the commons. But she cautioned against a rush to decentralization as a panacea. It could work only where the design principles applied. But even the design principles are no guarantee. Every set of rules is an experiment to be tested rather than a predefined solution. Decentralization was not a safeguard against vested interests, corruption of bureaucrats, elite takeovers, or devastation of the resource if the community and higher authorities do not carefully design the rules together.

"I wish to raise a strong cautionary note about the current brand of a decentralization craze that has swept through the development projects . . . there is no single blueprint for an effective organization to solve similar problems—let alone substantially different ones," Ostrom wrote in her later years as a warning against policies that too casually hand over control to users. "It is one thing to self-organize to create your own property and slowly develop the rules of association that enable a group to benefit from the long term management of that resource. It is quite something else to have a government tell you that now you have to manage something that the government can no longer handle itself!"

The successes of the residents of Torbël and Alanya and the Nepali farmers emerged slowly and incrementally as each group collectively tinkered with their rules over a long time. Nature has had an abundance of time to tinker with its rules. Birds flying in formation and ants carrying food to the nest have no blueprint or top-down instructions to get the job done. As the ant- and termite-like surveillance-containment method ultimately banished smallpox, the strategies of social insects conform with Ostrom's bottom-up challenge to Hardin's conventional wisdom about rules to govern the commons.

ANT TRAILS AND ZEBRA STRIPES ORGANIZE THEMSELVES

As long ago as the Bible's authors penned their words, people marveled at ants' collective ability to organize themselves unguided by commands from a higher authority. "Look to the ant, thou sluggard," they wrote, "consider her ways and be wise: which having no guide, overseer, or ruler, provideth her meat in the summer, and gathereth her food in the harvest."

Ants mesmerize all who watch what happens when a dropped crumb hits the floor. Ants first scurry randomly around the crumb. In a matter of seconds, the ants fall in line in army-like formation. They industriously march across the floor carrying pieces of the crumb as if someone were directing traffic. But there is no one coordinating or signaling directions. Each ant only follows her instinct to carry food to the nest. All she needs is her antennae to sense the ground beneath her.

Ants' fascinating ability to self-assemble in formation along the shortest path between food and their nest starts with a single ant and a crumb of food. One ant comes across a crumb, pulls off a piece to clutch in her jaws, and drags the food back to the nest. Along the way, the ant instinctively releases an odorless chemical that serves as a signpost to direct other ants to the same path. When a second ant follows the first ant's direction, she follows the same path to the crumb and leaves more signposts for more ants. A third ant follows, leaving more signposts to direct more ants as the chemicals laid by the first and second ants evaporate. The back-and-forth march between the crumb and the nest continues, with more ants following the pheromone signposts and leaving new signposts for more ants. The trail looks like someone had marshaled the ants to parade in unison. But each individual ant only follows the signposts right in front of her and lays down new ones. From a group of ants scampering in random directions, the straight line emerges. Food gets to the nest quicker and with less labor than if each ant searched for food on her own.

The pheromone-following strategy not only lines up the ants to march in unison. It steers them to the shortest path between the crumb and the nest. Imagine hikers blazing a new trail into unknown territory. Each hiker carries a bucket of blue paint to mark the path with blazes on trees. Sooner or later, the paint will wash off in the rain or fade in the sun. On a longer path, the blue blazes are likely to wash off before the hiker can paint new ones on a return trip. On a shorter path, the blazes have a better chance to get repainted from a returning hiker before they disappear. A second hiker choosing between the long and short path will likely follow the blazes on the short path but not the long one. In a self-reinforcing feedback, more short-path hikers will leave even more blazes to attract more hikers. On the longer path, the blazes will get washed away as fewer and fewer hikers take that trail. The paint-carrying hikers might take months or years to establish the route. In ants, blue blaze-like pheromones evaporate on the order of seconds. As more and more ants reinforce the short path, they abandon the long one. The process is simple, elegant, and beautiful—the ants collectively find the shortest path by each one following its individual genetically programmed instincts. The ants do not plan, remember, or see over the horizon to the crumb. They just follow the pheromone trail to decide which path to follow.

Back in the nest, the queen cannot control whether the worker ants take the long or short route to bring back food. Even if she could direct the workers, she has no way of knowing which way would be the best route. The very notion that the ant, whose role in the ant colony superorganism is solely to reproduce, deserves the "queen" label is a holdover from mid-eighteenth-century views about colonial-era monarchies. But alas, rather than the label's implied natural-world justification for servitude of native peoples, the queen has no power, authority, or ability to direct the workers. She is at the mercy of the worker ants who attack and destroy her if she doesn't produce enough offspring.

Besides thousands of species of ants, many other insects, birds, fish, and other animals live in city-like colonies, join forces to gather food and protect themselves, collectively build nests, or migrate in flocks. Canada Geese depend on one another to conserve their energy on the long annual flight to warmer climes. The spectacular V formation in the autumn sky cuts down on headwinds, allowing each goose to fly faster and farther in the wake of the bird in front. No single goose orders the others to fly in formation; the tight V shape emerges from a simple principle. Each goose positions itself to catch the updraft of air from the flapping wing of its neighbor in front. A school of fish with light reflecting from each one's shiny scales appears to dart in unison, as if one fish were prodding them from behind. The school emerges from each fish following simple rules: swim at the same speed as your neighbors, stay close, and avoid collisions.

Across the open savannas of Africa, Asia, and South America, spires of termite mounds can stand taller than a tall man. These sophisticated clay castles are architectural marvels. A single termite is no bigger than a grain of rice, but if termites were the size of humans their largest mounds would be a mile high and five miles wide. A tour inside a mound of particularly sophisticated African termites reveals brood chambers for young termites, fungus gardens, cooling vents, belowground tunnels, and a royal chamber for the queen and king. A single mound can house half a million individual termites.

Blind worker termites need to build rapidly to enlarge the royal chamber as the queen's abdomen swells with millions of eggs. The queen releases a pheromone so the workers know where to begin construction. One worker mixes a soil pellet with saliva and kneads it with its mandibles, then drops the pellet. The saliva has a pheromone that signals other workers to drop their pellets, which in turn signals other workers to drop their pellets at the same site. Once the worker adds its pellet to the pile, it returns to get more soil, guided by pheromones. Workers drop more and more pellets on separate building sites until they run out of building material. The result is a thick wall topped with spires to protect the queen. Without an architect to draw blueprints or a contractor to direct construction, the structure emerges as each individual termite follows its pheromone-following and pheromone-depositing instincts. The genetic coding to control these behaviors is a mystery we have yet to uncover.

One key feature ties together nature's intriguing array of ways that individual decisions benefit the group. Decisions come from the bottom up, in the Bible's words: with "no guide, overseer, or ruler." Individuals self-organize based on their perceptions of their local surroundings. That's the only way decisions to find the shortest path to food, locate a colony, or reduce headwinds can be accurate and timely. A queen termite in a nest or a Canada Goose at the head of the flock has little knowledge and no ability to perceive the local surroundings of each individual. Evolution favors decisions based on bottom-up information rather than top-down command from a central authority. Genetically encoded blueprints and step-by-step recipes can work for structures built by solitary individuals, such as spiders spinning webs. But a recipe can't work when the task requires collective action with interacting participants.

Evolution stumbled upon this relatively simple solution to organize behavior. Genetically code simple rules for individuals to follow based on their observations of their immediate surroundings. From the rules that each one follows, a pattern emerges without individual planning or knowledge. An initial self-reinforcing process, such as one ant's pheromones leading another ant to follow the same trail, ends when they collectively complete the task of taking the crumb to the nest.

The same process explains zebra stripes, leopard spots, and geometric patterns of colors on shells. The puzzle of how nature manages such a feat fascinated computer pioneer Alan Turing, who worked out the mathematics of the underlying biology. In a zebra embryo, black or white pigment cells in the skin pass into its growing hairs. Each black or white pigment cell diffuses chemicals across the zebra's skin like puddles around each cell. One chemical activates the pigment and another chemical inhibits. Black hairs dominate where a black activation signal overrides a black inhibition signal and white hairs dominate where a white activation signal overrides a white inhibition signal. Starting with a random distribution of black and white pigment cells, if the inhibition chemical diffuses faster and farther in one direction across the skin than the activation chemical, more hairs get black pigments close to other black hairs and white close to white—stripes. No genetic preprogramming of the entire pattern is required, only the genetic coding for the inhibition chemical to diffuse faster than the activation one. These self-organizing processes, prevalent in nature, run counter to ways human societies

often organize themselves with hierarchies, punitive laws, and centrally controlled management.

Despite what one might think of power grabbing and greed enabled by raw capitalism, Adam Smith saw the utility of a system driven by bottom-up, self-organizing individual decisions. His infamous "invisible hand" echoes evolution's insight that a central authority cannot replace local information perceived by individuals in the closest position to have accurate intelligence. Like shimmering fish schools, V-shaped formations, termite castles, and single-file ant marches that emerge unplanned from individuals following simple rules, economies grow from individual capitalists seeking profits for themselves. An individual "neither intends to promote the public interest, nor knows how much he is promoting it . . . he intends only his own gain, and he is in this, as in many other cases, led by an invisible hand to promote an end which was no part of his intention . . . By pursuing his own interest he frequently promotes that of the society more effectually than when he really intends to promote it." Adam Smith's "end" is not protection against predators or a royal chamber for a queen termite, rather it is the "distribution of the necessaries of life." Shopkeepers adjust their wares to what people buy. No central authority pre-plans how much flour, salt, or milk each shopkeeper should stock. The system runs itself.

One can question whether unfettered capitalism results in the most desirable distribution of basic necessities to the world's population. But indeed, the truth of Adam Smith's marketplace—and evolution's experience that bottom-up self-organization wins out over top-down command and control—has proven itself out over time as centrally planned economies mire in outdated technologies, waste, and inefficiencies.

Comparisons of ant colonies, flocks of geese, termite societies, and zebra stripes with capitalist economies and the communities of Torbël herders, Alanya fishers, and Nepali farmers cannot go too far. People do not follow instincts like pheromone-following ants or wing-flapping Canada Geese. Human-made rules and norms take the place of instincts in other species. Rules and norms are more fluid than genetically programmed rule-following behavior. Human societies devise their own rules about who can use how much of which resource. Every rule, regulation, law, or policy is an experiment in humanity's task to collectively organize ourselves.

Nature's rules provide a guide for civilization's experiments. Decentralized rules tailored to the ecology and culture of the place, developed and enforced by the users, tested over time, adaptable to changing conditions, and locally perceived as beneficial rather than punitive are more likely to endure than top-down directives. One-size-fits-all rules and scalable blueprints are less likely to ward off free riders and avoid tragedies. Top-down directives enable clockwork solutions. A central authority can deploy sensors in homes to cut down energy use, for example, or install detectors to keep people from speeding. In complex systems of interacting parts where the resource and the user continually change and adapt to each other, Ostrom's observations of successful self-organized communities ring true with nature's experience.

Ostrom studied small, mostly self-contained communities of fishers, herders, farmers, and forest-dwellers where local people could confront free riders in the coffeehouse. In today's interconnected world, for good or for bad, these communities are likely more connected with the outside world through markets and communications. Her design principles—clearly defined boundaries for users of the resource, ability of a community to make and enforce its own rules, and shared interests across the community—are increasingly less likely to be reality in many places. Particularly in an urban world with transient residents and no reason to invest time and energy to be part of a community, Ostrom's principles appear quaint. And in a world where vexing problems of climate change, pandemics, and international trade networks are global in scope, one can question whether the principles have any relevance.

Ostrom was well aware that her studies of small villages and fishing communities showed how people can self-organize to govern themselves, but they do not represent the interconnected, modern world where free riders can easily abuse the commons. Her eighth and final principle for rules to govern the commons made this point: when the resource is part of a larger system beyond the control of the users, build responsibility for making and enforcing rules at nested levels from local entities to international agreements. Writing about climate change, she cautioned that global solutions "negotiated at a global level, if not backed up by a variety of efforts at national, regional, and local levels, are not guaranteed to work well."

Lin Ostrom passed away on June 12, 2012, at the age of 78. Her close collaborator and husband Vincent died seventeen days later. Within four years of their deaths, countries around the world recognized the wisdom of her advice.

TOP-DOWN MEETS BOTTOM-UP AT THE NEGOTIATING TABLE

The atmosphere, the thin layer of gases that separates civilization from the black vastness of the universe, is ripe for tragedies. Billions of people use it to dump wastes from their tailpipes, factories, power plants, and burning fields and forests. No one owns the atmosphere, but everyone suffers the consequences of free-for-all dumping of greenhouse gases that swirl and mix in a global stew. No one has authority to make and enforce rules to govern how much each person, state, or country can dump into this common waste disposal. All these factors collide in a giant looming tragedy for a problem that scientists have known for over a century.

The Swedish physicist and chemist Svante Arrhenius as early as 1896 calculated that "the industrial development of our time" rivals geologic forces of weathering rocks as burned coal releases "carbonic acid," as he called carbon dioxide, into the atmosphere. The warming influence of a small quantity of greenhouse gases in the atmosphere, which re-radiate Earth's heat back to the surface, were not in question. The issue was what humanity's disruption to Earth's tried-and-true mechanisms for regulating the amount of carbon dioxide in the atmosphere means for civilization.

Arrhenius saw the upside: "By the influence of the increasing percentage of carbonic acid in the atmosphere, we may hope to enjoy ages with more equable and better climate especially as regards the colder regions of the earth. Ages when the earth will bring forth much more abundant crops than at present, for the benefit of rapidly propagating mankind."

More than a century hence, we know that Arrhenius was prescient in his analysis that "the industrial development of our time" is indeed an added player in the elegant cycling of carbon between oceans, atmosphere, and land. But the climate system and civilization are both so complex that simple conclusions about "benefit" don't hold. With myriad effects on rising seas along heavily populated coastlines, droughts and intense storms, crop failures, pests and diseases that thrive in the warmth, and vulnerable people unable to cope with the changes, the problem has mushroomed. "Humanity's period of grace," the last ten thousand years, when climate on our restless planet has been fairly stable and civilizations have thrived, is reaching an end.

December 12, 2015, an announcement from Paris hit newswires. Countries had been wrangling for decades over rules to stem the unceasing rise of greenhouse gases in the atmosphere. "With the sudden bang of a gavel Saturday night, representatives of 195 nations reached a landmark accord that will, for the first time, commit nearly every country to lowering planet-warming greenhouse gas emissions to help stave off the most drastic effects of climate change," read the article in the *New York Times*. "The deal, which was met with an eruption of cheers and ovations from thousands of delegates gathered from around the world, represents a historic breakthrough on an issue that has foiled decades of international efforts to address climate change."

The decades-long quest for an international agreement on climate change began on the heels of a prior, immensely successful treaty on the atmospheric commons. In 1987, countries collectively agreed to stop using industrial chemicals that waft into the air and destroy the life-enabling, thin layer of stratospheric ozone that protects all life from the sun's cell-destroying ultraviolet radiation. The Montreal Protocol on Substances that Deplete the Ozone Layer, so named for the city that hosted the convention, phased out production of the chemicals over a number of years. The problem turned out to be relatively simple with a clockwork solution. A ready substitute for the chemicals was in hand and the economic costs to businesses were low. The ozone layer began to recover, and Kofi Annan, head of the United Nations at the time, hailed the multicountry cooperation as "perhaps the single most successful international agreement."

Following this success, delegates to the Earth Summit in Rio de Janeiro adopted the United Nations Framework Convention on Climate Change in June 1992. The idea was "stabilization of greenhouse gas concentrations in the atmosphere at a level that would prevent dangerous interference with the climate system." Developed countries signed up to reduce their human-caused emissions of greenhouse gases to 1990 levels. In the interest of equity, fairness, and developing countries' concerns that the agreement would burden development goals, the 1990 target applied only to developed countries. The delegates left the details of implementation and enforcement to further discussions at a series of conferences to be held every year thereafter.

The climate agreement turned out to be anything but clockwork. The process was massively more complex than the one for the ozone layer. The coal-burning experiment that has powered civilization since before Arrhenius's time presents a seemingly unsolvable puzzle, unlike a synthetic chemical that industries can easily replace. If someone were to design a puzzle for countries to collectively solve, that person would be hard-pressed to invent a puzzle with more ill-fitting pieces than the climate conundrum. First, there is the sheer immensity of the costs to shift economies away from the coal, oil, and gas that have powered the engine of industrial development. Second, people don't see the need to pay attention to invisible gases and a seemingly distant and diffuse problem. Then there is the question of who is responsible for paying the costs: industrialized, developed countries whose historical emissions have driven their development and account for the lion's share of the problem; or industrializing countries whose development hinges on greenhouse-gas producing coal, gas, and oil? Shouldn't those responsible for the problem compensate those who suffer the consequences but played little role in creating it? But if those countries that are historically responsible for the problem curb their emissions, won't industrializing countries re-create the problem as they ramp up their emissions? What are the rules to assign how much of the atmospheric waste receptacle each country should be able to use? Should the rules allocate portions of the receptacle by a country's number of people, how much greenhouse gas the country emits for each person, how much the country emitted in the past, how much energy they require to develop, or any number of ways one could justify rules? And what rules will entice countries to comply in a world with essentially no way to hold free riders accountable? These were all questions left to the delegates of the conference of parties to sort out. They met in Berlin three years later and Geneva the year after.

In December 1997, delegates assembled for the third time in Kyoto. They aimed to pass targets and protocols that would reduce overall emissions of greenhouse gases by 5 percent by 2012. The protocol assigned targets to each country specifying how much it needed to reduce its emissions. Countries could trade their emission units and earn credit by reducing emissions in a developing country. Each country needed to report annually to a central secretariat in Bonn, Germany. The secretariat kept track of emissions units and provided instructions and guidelines for reporting. The procedures were top-down and centralized, as Hardin would have suggested.

Whether the Kyoto Protocol would have achieved its goal is impossible to know. The U.S. Senate didn't ratify the treaty based on objections that developing countries had no obligations to reduce their emissions. Indeed, in 2006 China took the reins from the United States as the country with the most emissions, although U.S. cumulative historical emissions as well as emissions per person remain the highest of any major country in the world. Without the United States, the Kyoto Protocol lost its teeth. Canada pulled out in 2011 because the two biggest emitters—China and the United States—were free riding without commitments. A coalition of developing countries including India, China, and Brazil called for compensation to help them cope with a problem that the developed countries created. Small island nations, including Fiji, Antigua and Barbuda, and Kiribati, formed an alliance to push for stricter targets as they watched their islands drown. The level of greenhouse gases in the atmosphere continued to climb.

A low point in the international negotiations hit in 2009, when delegates met for the fifteenth time in Copenhagen. Even President Obama's last-minute appearance couldn't break through the logjams. The talks collapsed in disarray without firm agreements about a way forward. But the failure laid the groundwork for cheers and ovations in Paris six years later. Only a drastic alternative to the top-down Kyoto protocol could revive the negotiations.

In the early days of the climate talks, Japan had suggested a "pledge-and-review" approach. Environmentalists derided the idea as "hedge and retreat." Decades later, a pledge-and-review approach turned out to be the winning strategy in Paris. Each country agreed to "prepare, communicate and maintain successive nationally determined contributions that it intends to achieve." Each country determines, based on its own situation, how much it can commit to reducing emissions. "Parties shall pursue domestic mitigation measures, with the aim of achieving the objectives of such contributions." Countries submit new plans every five years and "a Party may at any time adjust its existing national determined contribution with a view to enhancing its level of ambition." The fire-wall distinction between developed and developing countries no longer applied. No centrally determined targets or top-down goals guided the process.

After decades of delegates convening every year in cities around the world to devise rules for governing the commons, the wisdom of nature's strategies came to light. "Pledge and review" means each individual country develops its own plans based on its local circumstances, as birds join a flock or cells in zebra skins activate pigments. In an international community in which countries care about their reputations, an enhanced "level of ambition" acts like a self-reinforcing pheromone guiding ants to crumbs without top-down directives from a queen. Countries follow one another to avoid "name and shame," the equivalent of social pressure in the Alanya coffeehouse. And like those darting fish who only follow their own rules, the solution to climate will be greatest when countries see that reducing emissions fits their own local interests for whatever reason, either to partake in the renewable energy economy or to clean up their local environment. China's reduced rate at which emissions went up in recent years compared to the previous decade, for example, does not arise solely from concern for the climate. Coal-burning power plants rendered the air unbreathable and made citizens unhappy. Cleaning up dirty skies locally had a fortunate by-product—it reduced greenhouse gas emissions. In the words of Christiana Figueres, the effective United Nations diplomat who helped rebuild the climate negotiations after the collapse in Copenhagen, "None of them are doing this to save the planet," just as termites do not carry soil pellets with the good of the whole mound in mind. "The self-interest of every country is what is behind all of these measures."

The durability and success of the Paris Agreement is far from assured, particularly with political headwinds around the world. Even if all countries abide by their commitments, a very unlikely outcome, reductions in greenhouse gases are not sufficient to "prevent dangerous interference with the climate system." But Ostrom's no-one-size-fits-all principle to manage our shared global commons—to build responsibility for making and enforcing rules at nested levels from local entities to international agreements—has a chance for success where the top-down version failed.

In 2017, U.S. president Trump withdrew from the Paris Agreement, prioritizing the short-term interests of the dying fossil fuel industry over the planet's future. Responses from lower "nested levels" rose quickly and loudly. Thousands of mayors, governors, tribal leaders, businesses, and faith groups banded together to show support for the Paris Agreement under a "We Are Still In!" banner. More than a decade before the Paris meeting, the mayor of London convened his fellow mayors from forty of the world's largest cities to form a network to share information and take city-level actions against climate change. They pursued low-emission buses, green buildings, and plans to meet self-defined targets. Forty cities grew to more than ninety, from Accra to Beijing to Copenhagen all the way to Washington, D.C., in the effort still known as "C40."

What of the exponentially rising numbers of commons-gobbling hungry stomachs that propelled Hardin to write his influential essay in the late 1960s? It turns out he did not need to promote draconian solutions. He could have looked to nature's strategy. A global pattern emerges from individuals acting in their own interest, without oversight or coercion.

The average number of children born to each woman was already around two in the richer part of the world when Hardin wrote his article. Indeed, in poorer countries, the number hovered around six, as Hardin feared. In the 1970s that number started to fall. In South Korea, Mexico, Iran, and many countries around the world, people began to have fewer children. Targeted action, most notably China's one-child policy, and family planning were partly responsible. At least as influential were changing norms, women's education, the desire to have smaller families with more spent on each child, and better health care, generally what we think of as "development." With the right incentives and investments, self-interest and global concerns can align. The world is on track for Ehrlich's population bomb to fizzle by the middle of the century. Even so, all the additional people will need food, housing, and jobs, creating extra impetus to devise the right rules to govern common resources in our complex world.

The fourth and final piece of nature's advice draws on its experience with collective action when a task is too great for a single individual. Evolution elegantly settled on simple rules for each individual to follow. The solution—whether the task is to fly in formation, swim in a coordinated school, build a sophisticated castle, march in the shortest line, or grow hairs in a striped pattern—emerges as each individual follows rules based on local surroundings. No blueprint or masterplan is required. Cells and animals follow genetically preprogrammed rules that collectively give rise to zebra stripes, ant trails, and termites' architectural marvels. Humans, with ideas in the place of genes, devise their own rules and norms.

You can pattern your rules on nature's experience: each individual acts in self-interest, an individual makes decisions based on perception of the local environment, and no central authority issues directives. Evolution settled on this bottom-up solution to get around a basic problem. A queen ant or a bird at the head of the flock cannot effectively coordinate the actions of each and every individual in a complex matrix of individuals interacting with each other and with a changing environment.

Where top-down, clockwork solutions can solve your problems, use them. They are the easiest to deploy, as the alternative industrial chemical to replace the ozone-depleting one showed. If a new technology can save energy or cure a disease, by all means make it work. But don't fool yourself that clockwork, top-down solutions will work for all your problems. And be ready for seemingly top-down solutions to backfire, as Hardin's prescriptions did not stem the tragedies of the commons.

Over and over, you have learned that bottom-up solutions to your problems can work better than one-size-fits-all, top-down control. U.S. secretary of agriculture John Block's idea that all developing countries should import their food turned out to foment food riots, as Bill Clinton later conceded. The Nobel laureate Elinor Ostrom, through analysis of her meticulously-collected data, proved that people can organize to collectively manage their resources if given the freedom and information to make their own rules. Nepal recovered its forests when it returned management to local communities. Countries seeking to stem the rise in population learned that a citizenry with opportunities and education is a more effective contraceptive than draconian tactics. After decades of countries wrangling over top-down targets to keep climate in check, the power of bottom-up solutions came to light. And Asimov's fictitious Galactic Empire, patterned on the real Roman one, disintegrated as the emperor tried to govern from the center of the galaxy, counter to nature's advice.

As you make your rules, remember that people with eyes and ears on the ground and their own interests at stake can make more fitting decisions than a politician or bureaucrat in a far-off place. In a complex, unpredictable, rapidly changing world, let the people closest to a problem figure out the rules to manage their common resources. The process will be clumsy and messy, but the long-term benefit for enduring and adaptable rules and norms will pay off.

Nature's bottom-up strategy puts a spotlight on the meaning of leadership. True leaders don't lead through central commands. They have the wisdom to know the difference between a simple problem with a top-down solution and a complex one for which a clockwork solution won't suffice. They know they can set broad goals, but they can't dictate how to reach them with control-and-command tactics. When people need to share air and aquifers, or encounter other thorny problems in our complex world, a true leader empowers from the bottom up.

CYCLES OF RENEWAL

IN a fit of writer's block for this final chapter, I ventured downtown to the New York Public Library to see for myself the tiny Hunt-Lenox globe with medieval-style etchings of dragons and strange sea creatures. I was greeted by the iconic pair of bigger-than-life-sized marble lions that flank the imposing stairs to the grand building. The two majestic lions, known to New Yorkers by their names Patience and Fortitude, guard the entrance to stores of knowledge behind the arches fringed by palatial Corinthian columns.

Inside the library, I made my way to the collection of rare books and treasures on the third floor. The kind gentleman behind the desk explained that he could only show me a facsimile of the precious globe, as the actual globe is locked away in storage. The grapefruit-sized, reddish, copper copy of the globe revealed the fine lines and meticulous workmanship of the mapmaker. There they were, just below the equator, off the eastern coast of Asia, in uppercase block letters, the words "HC SVNT DRACONES."

The unnamed mapmaker must have known where to place the words of warning based on accounts of explorers returned from distant lands. Alongside the letters, etched pictures of dragons and monsters signaled seas and lands not yet seen by European eyes, although other peoples had lived in those lands for eons. Over time, as explorers ventured farther—with tragic outcomes for the people who had already laid claim to the land—European mapmakers more accurately traced the contours of continents and drew mountains and rivers in their actual locations. The mysteries of the unknown so beautifully depicted on the Hunt-Lenox globe lost their enigma over time.

The dragons on medieval maps indicate uncertainty, but the mapmakers could resolve the unknowns with more information. In complex adaptive systems—like the superorganism of modern civilization with billions of interacting people, ideas, and goods layered atop an equally complex and ever-changing planet—uncertain futures are not resolvable over time. They are fundamental, irresolvable features of the system. A serendipitous seemingly minor discovery can steer the course of civilization toward an unpredictable path. As one example from the previous pages, a perceptive milkmaid's eighteenth-century comment led to Jenner's vaccine, which led to Pasteur's discovery, which led to vaccination campaigns, saved lives, skyrocketing populations in the twentieth century, and ultimately to Hardin's alarm about the tragedy of the commons. As another example, the world-changing discovery that coal can replace animals, wind, and human muscles to power machines burgeoned into today's enormous climate change problem. Such path dependencies, when a small incident in the past amplifies into an enormous outcome for the future, are an unavoidable trait of complex adaptive systems.

In the most complex, long-lasting, and experienced of all adaptive systems, life on Earth, such path-dependent cascades occurred time and time again as the building blocks of life emerged from a primordial chemical stew, simple cells merged into multicellular organisms, and bacteria, reptiles, mammals and other life forms dominated and dwindled over billions of years. Path dependencies mean that no one can predict with precise accuracy what the future has in store. Dragons of uncertainty will always be with us, despite sophisticated models to predict climate trends over the coming decades and armies of analysts to predict the price of stocks in tomorrow's market. Predictions help us plan for the future, but even the best predictions cannot guard against an inherently unknowable future. Complex adaptive systems survive only if they can persist through an uncertain and unpredictable world.

Those who study the wonders of nature's complexity have highlighted its lessons for humanity for many decades. American ecologist Ward Allee called on international organizations to pay attention to this message in his 1951 book *Cooperation Among Animals with Human Implications*. He claimed that "the maintenance of smaller cooperative and competing units with the larger one is part of the scheme" in support of post–World War II cooperation among countries. In a 2008 article "Complex Systems: Ecology for Bankers," the brilliant physicist-turned-ecologist Robert May and colleagues Simon Levin and George Sugihara likened financial systems to ecosystems with a call to "identify conditions that dispose a system to be knocked from a seeming stability into another, less happy state." Marine ecologist Rafe Sagarin in his 2012 book *Learning from the Octopus* drew parallels between nature's resilience and adaptability needed for national security. No one could have predicted that a drunk driver would tragically strike and kill Sagarin on his bike just outside Biosphere 2, where the researcher conducted his experiments. Most likely, only a tiny fraction of diplomats, bankers, or national security experts ever knew about or paid heed to these arguments.

Analogies between nature and human civilization seem absurd on first blush. Nature has no empathy or concern for human values. Human societies take care of their sick, disabled, and nonproductive members. People pursue goals, both individually and collectively. Nature has no goals beyond each individual's innate aspiration to survive and reproduce. Civilizations evolve and adapt through ideas, rules, and norms. Nature evolves as individuals with the right combinations of genes survive long enough to reproduce.

The equivalence between life on Earth's struggles and the course of human civilization has its limits. No wonder the few who have called out the comparison get drowned by the rush toward simpler, clockwork-like solutions to manage businesses, cure diseases, and negotiate international agreements. Clockwork solutions seem efficient and decisive. But when people try to apply them to the global food trade, supply chains, the spread of pathogens, collective decisions to manage common resources, and other complex problems, seemingly efficient solutions can trigger ingrained lock-in and stagnation that leads to further problems for humanity.

Life on Earth and the interconnected superorganism of modern civilization—both complex adaptive systems—share some basic problems. To persist, they both need defenses against the irresolvable dragons of uncertainty that will always be part of their existence. They need to be able to recover from inevitable falls. They both need to move materials across dynamic networks or block the flow when materials carry danger. Both rely on collective actions from individuals working together. Ironically, the more civilization becomes sophisticated, urbanized, and seemingly removed from nature, the more it becomes interconnected and mired in complexity. Nature's strategies become even more relevant. They can postpone and cushion the fall in the endless cycle of growth, stagnation, breakdown, and renewal.

BREAKDOWN AND REBIRTH

I took the subway back uptown after spending a few hours in the New York City Public Library mesmerized by the Hunt-Lenox globe. The nearly unintelligible speaker system blasted announcements about delays in the lines. The walls of the Times Square subway stop were crumbling with age and rats scampered across the tracks. It was not surprising. Daily headlines report the poor management, outdated technologies, and dilapidated conditions of the New York City subway system. In one of the wealthiest cities in the world, in a system that transports millions of commuters every day, how could the city let the subway system fall into such a state?

Finally on the train to go uptown, I checked my email. One was from a student doing research in the village in India with the dusty bus stop and rickety red buses. The topic related to a patch of forest where a particularly aggressive and unwelcome weed had crowded out all the other trees and shrubs. Lantana, a noxious stubborn shrub originally from the Americas, forms dense thorny thickets. Other plants cannot grow where lantana invades. The tangle of thorns makes a walk through lantana nearly impossible. Villagers cannot collect firewood to cook or seeds to sell in the market. Removing lantana, like other invasive plants around the world that monopolize fields and pastures once they take root, is a tough task. Why can't native plants reclaim their ecosystems and shunt out the intruders?

Both the infrastructure of the New York City subway and the pesky weed in the forest became locked into their own complex systems. Once the city invested in building the subway over a hundred years ago, the expense to drastically change the technology and routes became too great to bear. The city makes incremental repairs rather than switching to a new, probably better system. In the forest, the land at some point must have suffered from overgrazing, fire, or some kind of disturbance. Lantana, which someone brought to India as an ornamental plant during British rule, got a foothold. It grew fast and crowded out other plants. Once it took root, the ecosystem couldn't recover. The forest patch became locked into a lantana future unless another drastic change occurs, which could happen when the villagers yank the troublesome weed out of the soil by its roots.

"Lock-in" results from path-dependent cascades in complex adaptive systems. It can occur in businesses and institutions, when norms become too embedded to easily change. Political lock-in happens all too often when leaders get into power and use it to control information and squelch opponents. "Carbon lock-in" tethers our energy systems to fossil fuels due to costs sunk in infrastructure and vested interest in the status quo. Even inferior technologies get locked in through incumbency and habit, like the QWERTY keyboard originally designed to keep typewriters from jamming but with little sense for computer keyboards. It's just too ingrained to change. Only a radical push can pull a system out of lock-in mode.

"Buzz" Holling, the Canadian ecologist whose writings resonate more with holistic Eastern philosophies than with reductionist Western science, sees lock-in as a feature of complex adaptive systems. He traces the cycle: growth, accumulation of knowledge and memory that eventually makes a system rigid and fragile, breakdown from an inevitable shock, and renewal to restart the cycle with new growth.

The cycle can occur as much in nonhuman systems as in human civilizations, though humans have more agency over the systems they control. In the complex system of many interacting trees, shrubs, animals, and microbes, a fresh, young forest grows fast and amasses biomass in its trunks, leaves, and branches. As time goes on, it becomes an old forest with towering, mature trees. The system becomes rigid. Nutrients stay stored in the trunks and leaves of the big trees. Only those trees that can tolerate shade have any chance to grow beneath the canopy. Until fire, wind, or some other disruption causes the system to reorganize, the system remains locked. For the system that governs the atmosphere of our unfortunate planetary neighbor Venus, once the air filled with volcanic gases and the planet's mantle hardened without plate tectonics to set off a self-regulating feedback as it did on Earth, water remained steam and stayed in the atmosphere. The planet became locked into a lifeless state without liquid water, and remains so in the absence of a drastic event such as a shove to a cooler location farther from the sun.

The cycle occurs in all complex systems from forests and planets to civilizations. Once a system becomes rigid, only a disruption—whether political upheaval, fire in a forest, or a crash to knock a planet out of its orbit—can shake a system loose. Then it can start anew and be flexible enough to withstand the next disruption. "Creative destruction," in the parlance of economist Joseph Schumpeter in the industrializing 1940s, can seem disastrous in a particular place and time, but it is necessary for a company or any other system to remain innovative, resilient, and endure over the long term.

If one is willing to take a very long view of past civilizations and put aside concerns for survival of our own, the waxing and waning of civilizations follow Holling's cycles. In its short tenure on the planet, human civilization has proven to be a remarkably resilient, creative, complex system that undergoes cycles of renewal. In the course of a single civilization, the system eventually becomes rigid. Society might keep investing in administration, irrigation systems, or other infrastructure to maintain sunk costs, even if the added benefit proves futile. Or authoritarian rulers can hold onto power despite opposition that eventually topples their regimes. Or people get stuck in their own worldviews, like Romulus Augustulus, the last emperor of Rome, or the fictional Cleon who ruled the Galactic Empire, and are blind to new ways that would help them adapt to shifting economies, invasions, or some other unexpected occurrence. Each civilization goes through the cycle of birth, growth, and rigidity, then reorganizes after authority and an organized structure break down. Another civilization can begin the cycle anew in a different time or place. Human civilization in one form or another persists.

American anthropologist Joseph Tainter holds the view that external disruption is not the only force that knocks back any individual civilization. "Society grows in complexity as a system,"

Tainter writes, meaning that bureaucracies flourish to manage armies, collect taxes, and build water systems, roads, and other public works to grow food and deliver services to citizens. When a problem arises, society invests in bureaucracy for a solution. "At some point in the evolution of society, continued investment in complexity as a problem-solving strategy yields a declining marginal return." In other words, as societies mature, bloated bureaucracies cost more than the benefits they deliver. Societies begin to crumble. Tainter sees this pattern in the collapse of civilizations from the Roman Empire to the Mayans to many other ancient societies. The decline comes slowly from within, making the civilization vulnerable and unable to harness the resources when an invasion, dry spell, or some other calamity strikes. Self-regulating mechanisms to winnow administrative costs are overwhelmed by the authorities' commitment to their administrative structures.

Through trial and error, civilization gains experience with new tools to persist through both self-inflicted and externally induced shocks. Shocks have inflicted damage to life on Earth for billions of years. Our brief experience is infinitesimal compared with nature's time-tested truths. With each crisis, humanity's tool kit gets fuller. We are learning that clockwork solutions are not the only tools in the tool kit.

AN EXPANDING TOOL KIT

Clockwork solutions have been part of humanity's tool kit ever since people started to use fire and tools millions of years ago. Machines and clockwork mechanisms have led to astounding benefits for humanity, from labor-saving devices to life-saving treatments to computers. Even in these pages, clockwork solutions play critical roles in solving problems. A technology to freeze-dry the smallpox vaccine was critical to Henderson's smallpox eradication effort. Synthetic substitutes for ozone-depleting chemicals made the Montreal Protocol on Substances that Deplete the Ozone Layer palatable to country delegates at the negotiating table. In the post–World War II technological bonanza of rockets, computers, communications, and medical procedures, the search for silver-bullet, quick-fix clockwork solutions has consumed the public imagination and the dollars devoted to research to find them. In the realm of biology, Sagarin reflected on the late-twentieth-century occupation with "focusing ever sharper at ever smaller scales of life" in a "precise world made of molecules." The same can be said for physics, medicine, and nearly every field. The whole gets lost in the parts.

In the clockwork world of the 1940s, a brave group of cutting-edge, renegade intellectuals ventured into the messy complexity of the interactions between humans, machines, and society. Psychologists, linguists, engineers, physicists, mathematicians, astronomers, biologists, anthropologists, political scientists, and economists convened, exchanged views, and debated in a series of conferences in New York. They met ten times between 1946 and 1953. The common goal was to decipher principles for society to manage and direct highly complex systems based on theories of how biological, social, and physical systems organize themselves.

The new field of cybernetics, derived from the Greek root *kubernetes*, meaning "steer" or "rudder," collapsed and disintegrated in the following decades. It was a misfit in a clockwork world of separate fields that scarcely interacted. In the words of one member of the Cybernetics Club, the famous anthropologist Margaret Mead who brought Samoan culture to the American consciousness, "We were impressed by the potential usefulness of a language sufficiently sophisticated to be used to solve human problems . . . It fascinated intellectuals and it looked for a while as if the ideas . . . would become a way of thought. But they didn't." The idea spread to the Soviet Union, where the government invested heavily in cybernetics on the premise that it would help manage their centrally planned economy. In the end, cybernetics didn't help much.

Despite the transience of cybernetics as a cohesive field, it laid a path to many useful tools. Alan Turing and others developed their ideas for computer science, artificial intelligence, and robotics, such as the Turing test to determine if a computer can "think" like a human. The cybernetic pioneer Norbert Wiener, who authored the book *Control and Communication in the Animal and Machine*, worked in electrical engineering on control mechanisms from thermostats to automated assembly lines, using principles of self-regulating feedbacks. Margaret Mead and others in the social sciences applied the principles to psychology, family therapy, and game theory and to understand how the brain operates the body and mind. The dynamic but poorly funded field of complexity science now carries the mantle of cybernetics. Its center of activity resides at an institute in the desert near Santa Fe whose mission is to explore the frontiers of complex systems.

In a path-dependent world, we cannot turn back the clock to know whether tools for managing highly complex systems would have emerged if cybernetics had persisted or if government and businesses had provided more funding for researchers. Perhaps epidemics could have spread less far, economic downturns could have been less costly, or conflicts could have been resolved sooner. Perhaps not. We do know that clockwork solutions alone will not stem the spread of new diseases, bring countries together to manage the global commons, or keep supply chains flowing in our twenty-first-century, complex, interconnected world.

When seemingly clockwork solutions mushroom into path-dependent complexity, the tool kit needs new tools to redress new problems that the solution created. Fleming's accidentally discovered penicillin derived from "mould juice" is a case in point. It saved countless lives. It seemed to be a magic bullet. Then the arms race between resistant bacteria and antibiotics began. Researchers scramble to keep one step ahead to provide humanity with life-saving antibiotics that became a mainstay of modern medicine. Borlaug's Green Revolution seemed to be the magic bullet to end hunger, and indeed it did eliminate the scourge of periodic famine. The unintended impacts ricocheted in the following decades into depleted groundwater, orphaned nutritious crops, and an overabundance of calories. Humanity is grasping for solutions to these problems.

Even without university programs that prepare leaders and managers for a complex, uncertain world, humanity's collective intelligence is learning how to persist in an unknowable future. With each crisis new tools emerge. Each tool is an experiment to be tested as more crises occur. As narrated in previous chapters, business leaders, public officials, investors, and planners are stumbling upon ways to reduce their risks and weather the crises of uncertain times. They are finding these strategies not because ecologists advise them to do so, but because they figured out through trial and error that the strategies work. Probably inadvertently, without any awareness that evolution has settled on similar solutions over time, humanity continually expands its tool kit.

One problem for both modern civilization and nature, as chapter 2 highlights, is how to keep conditions from spinning out of control. Human societies can only avoid breakdown if they can steady themselves after a calamity, the way the grand cycling of carbon in the atmosphere regulates climate and shivering and sweating regulate body temperature. Human ingenuity has crafted many tools that emulate nature's elegant strategies. Some are mechanical, such as the contraption that the inventor of the steam engine, James Watt, devised to solve his problem. His engine sped up too fast with too much steam from the boiler and slowed down with too little steam. Watts had a clever solution to keep the speed constant. He mounted a pair of metal balls on a vertical shaft. When the engine's speed increased, the metal balls spread apart like a skirt of a ballroom dancer. The spreading balls operated a valve that reduced the flow of steam into the engine. When the engine slowed down, the balls moved closer to the rod which opened the valve to let in more steam. The speed stayed fairly constant with no human operator. The flyball governor works through feedback control, like the temperature-sensitive weathering of carbon from rocks and the many self-regulating mechanisms in nature.

Some of humanity's tools simulate nature's time-tested control feedbacks, as did Watt's flyball governor and as circuit breakers to halt the stock market from sliding into free fall do today. Other controls are not so obvious, like the self-regulating role of small forest fires to keep a big fire from blazing out of control. Frederick Erskine Olmsted's successors had to unlearn the forest management policies Olmsted had imported from Germany. But after one hundred years of the 10 A.M. policy, the U.S. Forest Service reversed course. Prescribed burns to manage forests are now matter of policy.

Another problem that both life on Earth and human civilization face: reserves to recover after a fall to make renewal in Buzz Holling's cycle possible. Chapter 3 features diversity as the strategy for this problem. Earth could have become a lifeless planet without its rich stores of diversity each time an asteroid hit, the climate became too hot or cold, or another potential catastrophe struck. Species would have been unable to adapt without diversity in the genetic makeup among individuals. Civilizations rely on this same diversity of species to produce food, its most critical function. The most valuable banks in the world do not hold jewels or gold, as the Syrian wild grass that saved the wheat crop in the Midwest of the U.S. shows. The thousands of underfunded, unappreciated gene banks around the world, ranging from small communities of farmers who save their seeds to imposing official government buildings in Fort Collins and St. Petersburg, are truly the ultimate tool for humanity's pantry to recover and adapt.

What plants and animals store in their genes, humanity stores in its languages and its traditions—knowledge and memory of strategies to survive through uncertainties. The science fiction writer Isaac Asimov, who had the wisdom to portray the fictional Galactic Empire as a complex adaptive system, knew the value of stored knowledge. Asimov gave Hari Seldon, the central character in the many volumes of the Foundation series, the foresight to collect all knowledge accumulated across the galaxy in an *Encyclopedia Galactica* as the empire slid into self-destruction. Hari settled a group of people to assemble the encyclopedia on a small planet at the extreme edge of the galaxy. The knowledge, or rather the process of compiling the knowledge, shortened the dark ages after the collapse from tens of thousands to one thousand years in Asimov's imaginary future.

Humanity's tool kit to recover after a fall depends on storehouses in the forms of genes and knowledge accumulated over many millennia to garner resources and organize society. Science is unraveling the power of biological diversity and its value to humanity, in the human gut and in ecosystems that provide food, water, and material needs. In the meantime, bulldozers and pavement are eliminating the storehouses of biological diversity and removing tools from the tool kit. And as languages disappear with the death of the last elders and modernity dilutes cultures around the world, the storehouse of knowledge diminishes. Gene banks, places safe from human exploitation, and efforts to pass language and knowledge to the next generation are all essential tools in humanity's tool kit to persist through uncertain shifts in climate, diseases, and economic upheavals.

Yet another problem that both life on Earth and modern civilization share is the yin-yang necessity and danger of networks, as chapter 4 notes. Water, blood, and sugars travel through veins in biological systems. Food, goods, people, and ideas move through transport systems in human civilization. Some networks need to function even if one part gets damaged, like the water- and sugar-carrying veins of a leaf. Other networks need to prevent dangerous substances from spreading, like the social networks that can transmit disease from one insect to another. The architecture of networks in nature fit the purpose—loopy in leaves and modular in ants' nests.

In modern civilization, hub-and-spoke, small-world networks tend to develop in the World Wide Web, supply chains, trade routes, and all kinds of human-constructed networks. The hubs with the most traffic attract connections in an unequal, rich-get-richer race to attract new nodes. Hub-and-spoke networks convey goods efficiently until a blockage knocks out a hub for one reason or another. The Thailand flood that crippled the global hard disk market for months; the eventual success of Paul Baran's rejected leaf-type, decentralized network as the basis for the internet; and Bill Clinton's public acknowledgment of John Block's bad advice to developing countries to import all their staple foods added to humanity's collective knowledge about the downsides of hub-and-spoke networks. When food needs to move across networks to keep people fed and communications need to reach their destination, a leaf's loopy network might be the better model.

Don Henderson's and William Foege's serendipitously discovered strategy to stop the spread of the deadly smallpox virus only came to light when vaccine supplies ran short. Without enough vaccines to go around, the fallback strategy was to vaccinate the victim's contacts rather than nearly everyone. By forming a ring of immunity around the victim, the new strategy worked better. It proved the point that the ant's modular network structure efficiently cuts off the spread in human society as well. The 2020 scramble to contain the world-changing pandemic of a novel, highly contagious coronavirus reinforces that wisdom: take quick action to stem the spread before the infection can spread among communities, cities, and countries. If society manages the structure of its networks to maximize their benefits and minimize their risks, the toolbox has more tools to manage potential catastrophes.

Finally, no society could function without individuals working together to achieve a common goal, the topic of chapter 5. Ants gather food collectively. Termites construct their mounds. Fish swim in schools to protect against predators. Birds fly in flocks to combat headwinds. Humans are the most social and collaborative creatures of all. As instinct and genes govern the behavior of insects and other animals, rules and norms guide human behavior. Which rules and norms can keep herdsmen's cattle from overgrazing the commons, fishers from depleting the stock, or countries from allowing their power plants and industries to spew dangerous gases into the atmosphere? Hardin argued for top-down rules imposed from a central authority to overcome the tragedy of the commons. Ostrom countered that people manage the commons, fish stocks, and other common resources better with bottom-up rules based on local knowledge and devised by those who need to follow them.

The idea that one set of rules does not fit all, without a central commanding authority, seems discordant to the human psyche.

It runs counter to the popular notion of "scalability." But evolution's experience shows that order can emerge without blueprints or control from a top-down authority. Countries negotiating the rules for collectively managing the atmosphere learned from failure that evolution's bottom-up solution to collective action can apply to people as well. Once again, humanity's tool kit for organizing itself expanded beyond Hardin's top-down directives.

Unwittingly and paradoxically, modern civilization is learning to solve these four types of problems to buffer against unpredictable futures as nature has done for millions of years: keeping conditions from spinning out of control; stocking reserves to recover and adapt; managing the benefits and dangers from networks; and organizing collective action for the common good. With each devastating fire, crop-killing disease, shortage of vaccines, collapse of international negotiations, epidemic of a deadly disease, or other crisis, new tools emerge that provide the possibility for less harm the next time a crisis occurs. More tools will emerge as civilization faces new crises. The looming question is whether the tools are effective enough or whether society can put them into practice soon enough to recover from the crises that are in store.

As our uber-interconnected modern civilization and a changing planet intersect and collide, the world has never witnessed a system so complex, dynamic, and ripe with potential and peril. In the very long view, modern civilization, like its predecessors, will likely follow the cycle of growth, rigid stagnation, disruption, and renewal. Through humility and attentiveness to life on Earth's long experience with persisting through uncertainty, stagnation can be less stagnant, disruption less disruptive, and renewal more rapid.

ACKNOWLEDGMENTS

THIS book was a long time in gestation and went through many iterations. Many people were part of the journey and to them I am grateful. Jit Bajpai, Terry Chapin, Winslow Hansen, Tamar Haspel, Simon Levin, and Jane Nisselson provided helpful comments on the manuscript. Illustrator Sophie Capshaw-Mack was delightful to work with on the drawings. Elizabeth Evans gave pivotal direction on earlier attempts at shaping the ideas into a manuscript. David Bowman, Joel Cracraft, Maria Diuk-Wasser, David Helfand, Shahid Naeem, Harini Nagendra, Michael Puma, Jeff Shaman, Melinda Smale, and Duncan Watts were kind enough to review chapters for factual errors. Miranda Martin and the staff at Columbia University Press managed the editing and production with utmost professionalism.

The ideas and stories in this book are only possible because many people have devoted their careers to swimming upstream against the currents of a clockwork-world. Only some are mentioned by name in this book but I am grateful to them all. Inevitably, some mistakes still lurk in these pages, for which I alone am responsible.

Most importantly, without the inspiration from family—Jit, Triveni, Erik, Avi, Yami, and Asha—and extended family and friends too numerous to list, nothing else matters.

NOTES

PROLOGUE

iii “This is not a drill.” Director-general of the World Health Organization Dr. Tedros Adhanom Ghebreyesus gave this warning on March 6, 2020 (reported by CNN).“WHO Chief on Coronavirus: This Is Not a Drill,” *CNN World*, https://www.cnn.com/videos/world/2020/03/06/who-coronavirus-director-general-this-is-not-a-drill-sot-intl-hnk-vpx.cnn.

1. THE DRAGONS ARE BACK

2 riches lay beyond: Shirin Elahi, “Here Be Dragons . . . Exploring the ‘Unknown Unknowns,’ ” *Futures* 43, no. 2 (2011), discusses the Hunt-Lenox globe in the New York Public Library and how the current lack of knowledge is analogous to symbols on medieval maps.

3 capacity to cope: The uncertainties in climate predictions are discussed in R. DeFries et al., “The Missing Economic Risks in Assessments of Climate Change Impacts” (Policy Insight, Grantham Research Institute on Climate Change and the Environment, London, September 2019).

5 years hence: Asimov imagined that the Galactic Era began in 10,000 CE. Hari Seldon meets Cleon I on Trantor in year 12,000 Galactic Era, suggesting that the novel takes place in 22,000 CE. http://asimov.wikia.com/wiki/Asimov_Timeline.

5 their bra straps: Asimov's behavior with women is described in A. Nevala-Lee, "Asimov's Empire, Asimov's Wall," *Public Books*, January 7, 2020.

6 disposing of wastes: Michael White, *Isaac Asimov: A Life of the Grand Master of Science Fiction* (New York: Carroll & Graf, 2005).

6 a fictional future: Joseph F. Patrouch, *The Science Fiction of Isaac Asimov* (New York: Doubleday, 1974).

6 Odoacer: Bryan Ward-Perkins, *The Fall of Rome: And the End of Civilization* (New York: Oxford University Press, 2006).

6 read and write: Ward-Perkins, *The Fall of Rome*; Donald A. Yerxa, "An Interview with Bryan Ward-Perkins on the Fall of Rome," *Historically Speaking* 7, no. 4 (2006): 31–33.

7 layer upon layer of ice: Joseph R. McConnell et al., "Lead Pollution Recorded in Greenland Ice Indicates European Emissions Tracked Plagues, Wars, and Imperial Expansion During Antiquity," *Proceedings of the National Academy of Sciences* 115, no. 22 (2018): 5726–31.

7 less common: Ward-Perkins, *The Fall of Rome*.

7 sophisticated society: Many books have been written about the fall of the Roman Empire, most famously the six volumes by Edward Gibbon, *The History of the Decline and Fall of the Roman Empire* ([London?]: W. Strahan and T. Cadell, 1776). The view adopted by historians has shifted from portraying the decline as a barbaric conquest to a change in culture. See Ward-Perkins, *The Fall of Rome*; Niall Ferguson, in "Complexity and Collapse: Empires on the Edge of Chaos," *Foreign Affairs* 89, no. 2 (2010): 18–32, discusses the collapse of civilizations more generally as the dynamics of complex systems.

7 underlying reasons: The 210 reasons for the decline of the Roman Empire were compiled by Alexander Demandt, *Der Fall Roms* (Munich, Germany: Beck, 1984), and translated in Karl Galinsky, *Classical and Modern Interactions: Postmodern Architecture, Multiculturalism, Decline, and Other Issues* (Austin: University of Texas Press, 1992).

8 "less adaptable to change": Ward-Perkins, *The Fall of Rome*, 158

8 tax revenue: Preconditions and triggers for historic societal collapses are discussed in Karl W. Butzer, "Collapse, Environment, and Society," *Proceedings of the National Academy of Sciences* 109, no. 10 (2012): 3632–39.

8 resisted change: N. Dunning, T. Beach, and S. Luzzadder-Beach, "Kax and Kol: Collapse and Resilience in Lowland Maya Civilization," *Proceedings of the National Academy of Sciences* 109, no. 10 (2012): 3652–57.

9 downfall inevitably occurred: Joseph Tainter, in *The Collapse of Complex Societies* (Cambridge: Cambridge University Press, 1988) and "Social Complexity and Sustainability," *Ecological Complexity* 3, no. 2 (2006): 91–103, argues persuasively that civilizations decline when the cost of maintaining the bureaucracy exceeds the marginal return from those costs.

9 "degree of humility": Yerxa, "An Interview with Bryan Ward-Perkins on the Fall of Rome," 3.

10 unpredictable complexity: A readable account of the study of the origin of life can be found at: www.bbc.com/earth/story/20161026-the-secret-of-how-life-on-earth-began.

10 into seawater: The physiology and habitat of the last universal common ancestor is described in Madeline C. Weiss et al., "The Physiology and Habitat of the Last Universal Common Ancestor," *Nature Microbiology* 1, no. 9 (2016): 16116.

11 its disguises: Kevin Kelly, *Out of Control: The New Biology of Machines, Social Systems, and the Economic World* (London: Hachette UK, 2009), 67.

11 psychedelic paisley: The originator of the riddle was Stewart Brandt, author of the *Whole Earth Catalogue* in the 1970s. The other responses from experts are described in Kelly, *Out of Control.*

12 on its own: See David M. Althoff, Kari A. Segraves, and Marc T.J. Johnson, "Testing for Coevolutionary Diversification: Linking Pattern with Process," *Trends in Ecology & Evolution* 29, no. 2 (2014): 82–89, for coevolution of pollination.

13 mammals to dominate: Summarized from chapter 2 in R. DeFries, *The Big Ratchet: How Humanity Thrives in the Face of Natural Crisis* (New York: Basic Books, 2014).

14 control it: See Yaneer Bar-Yam, "Complexity Rising: From Human Beings to Human Civilization, a Complexity Profile," (2002), and David Ronfeldt, "In Search of How Societies Work," *RAND Pardee* (2007), for discussion of evolution in the organization of societies.

2. RECOVERY FROM A CRASH: INSTALL ALL CIRCUIT BREAKERS

19 circuit breaker–triggering rules: The Flash Crash is described in Eric Aldrich, Joseph Grundfest, and Gregory Laughlin, "The Flash Crash: A New Deconstruction" (SSRN, March 26, 2017); Albert J. Menkveld

and Bart Yueshen, "The Flash Crash: A Cautionary Tale About Highly Fragmented Markets," *Management Science* 65 (2019): 4470–4488; Ali N. Akansu, "The Flash Crash: A Review," *Journal of Capital Markets Studies* 1, no. 1 (2017): 89–100; Andrei Kirilenko et al., "The Flash Crash: High-Frequency Trading in an Electronic Market," *Journal of Finance* 72, no. 3 (2017): 967–98. The U.S. Securities and Exchange Commission and the U.S. Commodity Futures Trading Commission analyzed the event in a detailed report. U.S. Securities Exchange Commission, "Findings Regarding the Market Events of May 6, 2010: Report to the Staffs of the Cftc and Sec to the Joint Advisory Committee on Emerging Regulatory Issues," (New York: U.S. Securities Exchange Commission, 2010). In 2016, a London-based trader named Navinder Singh Sarao pleaded guilty to spoofing the market on the day of the Flash Crash, although it is not clear that his actions actually caused the crash. The history of circuit breakers is described in Avanidhar Subrahmanyam, "Algorithmic Trading, the Flash Crash, and Coordinated Circuit Breakers," *Borsa Istanbul Review* 13, no. 3 (2013): 4–9; and J. Moser, "Circuit Breakers," *Economic Perspectives* 14, no. 5 (September 1990).

19 coronavirus-induced panic: Quote from the president of the New York Stock Exchange is from Stacey Cunningham in J. Cox and M. Bloom, "The Market Triggered a 'Circuit Breaker' That Kept Stocks from Falling Through the Floor. Here's What You Need to Know," *CNBC*, March 9, 2020. Information about March 12, 2020 is from T. Heath, T. Telford, and H. Long, "Dow Plunges 10 Percent Despite Fed Lifeline as Coronavirus Panic Grips Investors," *Washington Post*, March 12, 2020.

23 lifeless planets: The carbonate–silicate cycle was first described in James C. G. Walker, P. B. Hays, and James F. Kasting, "A Negative Feedback Mechanism for the Long-Term Stabilization of Earth's Surface Temperature," *Journal of Geophysical Research: Oceans* 86, no. C10 (1981): 9776–82. Plate tectonics is described in S. Ross Taylor and Scott M. McLennan, "The Evolution of Continental Crust," *Scientific American* 274, no. 1 (1996): 76–81. Other sources on the carbon cycle are Tim Lenton, Andrew Watson, and Andrew J Watson, *Revolutions That Made the Earth* (Oxford: Oxford University Press, 2011); and Vaclav Smil, *Cycles of Life: Civilization and the Biosphere*, Scientific American Library (New York: Freeman, 1996).

23 animals: S. Blair Hedges et al., "A Molecular Timescale of Eukaryote Evolution and the Rise of Complex Multicellular Life," *BMC*

Evolutionary Biology 4, no. 1 (2004): 2, discuss the origins of animal life and the role of oxygen and mitochondria. Daniel B. Mills et al., "Oxygen Requirements of the Earliest Animals," *Proceedings of the National Academy of Sciences* 111, no. 11 (2014): 4168–72, points to the low-oxygen conditions for the earliest forms of animal life and questions the primary role of oxygen.

24 shortcomings of warm-bloodedness: Frank Seebacher, "The Evolution of Metabolic Regulation in Animals," *Comparative Biochemistry and Physiology Part B: Biochemistry and Molecular Biology* 224 (2017): 195–203, provides several examples of self-regulating mechanisms in humans.

25 insulin's circuit-breaking role: A clear description of the role of insulin and glucagon is www.endocrineweb.com/conditions/diabetes/normal-regulation-blood-glucose. The link between sugar intake and type 2 diabetes is reviewed in Emily Sonestedt et al., "Does High Sugar Consumption Exacerbate Cardiometabolic Risk Factors and Increase the Risk of Type 2 Diabetes and Cardiovascular Disease?," *Food & Nutrition Research* 56, no. 1 (2012): 19104; and Sanjay Basu et al., "The Relationship of Sugar to Population-Level Diabetes Prevalence: An Econometric Analysis of Repeated Cross-Sectional Data," *PLoS ONE* 8, no. 2 (2013): e57873. Some sources indicate that high sugar diet is not directly responsible for type 2 diabetes. See the following for a clear explanation of current understanding: www.pcrm.org/nbBlog/does-sugar-cause-diabetes.

27 saved the moonwalk: Margaret Hamilton's contribution to the moon landing is described in Robert McMillan, "Her Code Got Humans on the Moon—and Invented Software Itself," *Wired*, October 13, 2015, https://www.wired.com/2015/10/margaret-hamilton-nasa-apollo; and her own description appears in Margaret H. Hamilton and William R. Hackler, "Universal Systems Language: Lessons Learned from Apollo," *Computer* 41, no. 12 (2008): 34–43.

27 a single year: Mark Nelson et al., "Closed Ecological Systems, Space Life Support and Biospherics," in *Environmental Biotechnology*, ed. Lawrence K. Wang, Volodymyr Ivanov, and Joo-Hwa Tay, 517–65, *Handbook of Environmental Engineering* 10 (New York: Springer, 2010); and Mark Nelson and William F. Dempster, "Living in Space—Results from Biosphere 2's Initial Closure, an Early Testbed for Closed Ecological Systems on Mars," in *Strategies for Mars: A Guide to Human*

Exploration, ed. American Astronautical Society, Carol R. Stoker, and Carter Emmart, 363–90, Science and Technology Series 86 (San Diego, Calif.: Univelt, 1996).

27 into space: David P. D. Munns and Kärin Nickelsen, "To Live Among the Stars: Artificial Environments in the Early Space Age," *History and Technology* 33, no. 3 (2017): 272–99, references the cost of glass of water in space.

27 "other solar systems": Quotes from Hawkins are from H. Wall, "Stephen Hawking Never Reached Space, but He Sought to Lift All of Humanity," *Space.com*, March 14, 2018.

28 no bigger than a house: Nelson et al., "Closed Ecological Systems," describes the history of experiments with closed systems.

28 Bass hired Allen as director: G. de Lama, "Biosphere 2 Proves a Hothouse for Trouble: Project Yields a Crop of Rivalry, Confusion," *Chicago Tribune*, April 16, 1994.

29 two-year experiment: Nelson and Dempster, "Living in Space"; and Nelson et al., "Closed Ecological Systems."

29 "sophisticated telecommunications": Quotes are from John C. Avise, "The Real Message from Biosphere 2," *Conservation Biology* 8, no. 2 (1994): 328

30 eighty plants: Nelson et al., "Closed Ecological Systems"; and Mark Nelson, William F. Dempster, and John P. Allen, "Key Ecological Challenges for Closed Systems Facilities," *Advances in Space Research* 52, no. 1 (2013): 86–96, describe diets in Biosphere2.

30 their crops: Nelson and Dempster, "Living in Space."

30 guinea pigs: Munns and Nickelsen, "To Live Among the Stars," discuss a calorie-restricted diet

30 Mt. Everest: Mount Everest is 29,000 feet. Equivalent elevation of Biosphere2 was 17,500 feet when oxygen dropped from 21 percent to 14 percent. Joel E. Cohen and David Tilman, "Biosphere 2 and Biodiversity—the Lessons So Far," *Science* 274, no. 5290 (1996): 1150–51.

31 Biosphere 2: Jordan Fisher Smith, "Life Under the Bubble," *Discover*, December 19, 2010.

31 locking up oxygen in the structure: Jeffrey P. Severinghaus et al., "Oxygen Loss in Biosphere 2," *EOS, Transactions, American Geophysical Union* 75, no. 3 (1994): 33–37.

31 plants could not reproduce: J. P. Allen, Mark Nelson, and Abigail Alling, "The Legacy of Biosphere 2 for the Study of Biospherics and

Closed Ecological Systems," *Advances in Space Research* 31, no. 7 (2003): 1629–39; and Cohen and Tilman, "Biosphere 2 and Biodiversity."

31 they didn't all get along: Allen, Nelson, and Alling, "The Legacy of Biosphere 2"; Mark Nelson, William F Dempster, and John P Allen, "Key Ecological Challenges for Closed Systems Facilities"; and Smith, "Life Under the Bubble."

32 a restraining order against management: Smith, "Life under the Bubble."

33 "the new situation": Quote from a letter from Abigail Alling to Donella Meadows reported in http://donellameadows.org/archives/biosphere-2-teaches-us-another-lesson.

33 Alling said in defense: S. Cole, "The Strange History of Steve Bannon and the Biosphere2 Experiment," *Motherboard*, November 15, 2016.

33 "two years": Cohen and Tilman, "Biosphere 2 and Biodiversity," 1150.

35 release their seeds: Sources for the section on fire and Smokey Bear are: A. Paige Fischer et al., "Wildfire Risk as a Socioecological Pathology," *Frontiers in Ecology and the Environment* 14, no. 5 (2016): 276–84; Stefan H. Doerr and Cristina Santín, "Global Trends in Wildfire and Its Impacts: Perceptions Versus Realities in a Changing World," *Philosophical Transactions of the Royal Society of London B: Biological Sciences* 371, no. 1696 (2016): 20150345; Sean A. Parks et al., "Wildland Fire as a Self-Regulating Mechanism: The Role of Previous Burns and Weather in Limiting Fire Progression," *Ecological Applications* 25, no. 6 (2015): 1478–92; Geoffrey H. Donovan and Thomas C. Brown, "Be Careful What You Wish For: The Legacy of Smokey Bear," *Frontiers in Ecology and the Environment* 5, no. 2 (2007): 73–79; Roberta Robin Dods, "The Death of Smokey Bear: The Ecodisaster Myth and Forest Management Practices in Prehistoric North America," *World Archaeology* 33, no. 3 (2002): 475–87; Jesse Minor and Geoffrey A. Boyce, "Smokey Bear and the Pyropolitics of United States Forest Governance," *Political Geography* 62 (2018): 79–93; Philip N. Omi, "Theory and Practice of Wildland Fuels Management," *Current Forestry Reports* 1, no. 2 (2015): 100–17.

35 "New Mexico": L. Charlton, "Smokey Bear Dies in Retirement," *New York Times*, November 10, 1976.

37 a big, intense fire: Fred Cahir et al., "Winda Lingo Parugoneit or Why Set the Bush [on] Fire? Fire and Victorian Aboriginal People on the Colonial Frontier," *Australian Historical Studies* 47, no. 2 (2016): 225–40;

and Rhys Jones, "Fire-Stick Farming," *Fire Ecology* 8, no. 3 (2012): 3–8. There is some controversy about the extent of "fire stick farming" by the Aborigines, but little doubt that fire was important in the Aboriginal economy. Alan N. Williams et al., "Exploring the Relationship Between Aboriginal Population Indices and Fire in Australia over the Last 20,000 Years," *Palaeogeography, Palaeoclimatology, Palaeoecology* 432 (2015): 49–57.

37 "burning, burning, ever burning": Quoted in Stephen J. Pyne, "Firestick History," *The Journal of American History* 76, no. 4 (1990): 1135.

37 the Forest Service: "Frederick Erskine Olmsted," *Journal of Forestry* 23, no. 4 (1925): 338–39.

37 "keep fire out absolutely": Quote from Minor and Boyce, "Smokey Bear and the Pyropolitics of United States Forest Governance," 82.

37 "these ecosystems": Quote from Dods, "The Death of Smokey Bear," 479.

38 "a commodity belonging to us": Aldo Leopold, *A Sand County Almanac, and Sketches Here and There*, Outdoor Essays & Reflections (New York: Oxford University Press, 1989), viii.

38 "natural ground fires": Aldo Starker Leopold, *Wildlife Management in the National Parks* (U.S. National Park Service, 1963), 6. The report is available at https://www.nps.gov/parkhistory/online_books/admin_policies/policy4-leopold.htm.

38 extinguish all fires: Leopold, *Wildlife Management in the National Parks*, 6

38 in 1968: Hal K. Rothman, *A Test of Adversity and Strength: Wildland Fire in the National Park System* (Washington, D.C.: National Park Service, 2005), 134.

38 "grapple with that idea": Quote from interview with National Park Service researcher Jan van Wagtendonk on June 13, 2002, reported in Rothman, *A Test of Adversity and Strength*, 136.

38 prescribed fires to burn: Rothman, *A Test of Adversity and Strength*, 158.

40 "renewal of the park ecosystems": Quote from Rothman, *A Test of Adversity and Strength*, 199.

40 be rescinded: Sheila Olmstead, Carolyn Kousky, and Roger Sedjo, "Wildland Fire Suppression and Land Development in the Wildland/Urban Interface" (Washington, D.C.: Resources for the Future, 2012).

40 "Colorado Blaze Continues to Grow": Sources for headlines, respectively: I. Kwai, "Devastating Australia Bush Fire Destroys Scores of Homes," *New York Times*, March 19, 2018; "Australia'a Worst-Ever Wildfires Kill 130," *CBS News*, February 8, 2009; J. Medina, L. Stack, and J. Bromwich, "Tens

of Thousands Evacuate as Southern California Fires Spread," *New York Times*, December 5, 2017; P. Wright, "Colorado Blaze Continues to Grow; Entire San Juan National Forest to Close," *The Weather Channel* 2018.

41 California's history: A. Reyes-Velarde, "California's Camp Fire Was the Costliest Global Disaster Last Year, Insurance Report Shows," *Los Angeles Times*, January 11, 2019.

41 evacuate their homes: J. Bates et al., "Firefighters Make Progress Against Fires Raging in California," *Time*, October 25, 2019.

41 "global warming": Quote from Park Williams in K. Stark, "Climate Change Is Driving California Wildfires. The Kincaid Fire? No So Much," *KQED Science*, November 6, 2019.

42 ignite more fires: I. Penn, P. Eavis, and J. Glanz, "How Pg&E Ignored Fire Risks in Favor of Profits," *New York Times*, March 19, 2019.

42 a tinderbox: Rachael H. Nolan et al., "Causes and Consequences of Eastern Australia's 2019–20 Season of Mega-Fires," *Global Change Biology* 26, no. 3 (2020): 1039–1041.

42 circumnavigated the planet: B. Resnick, U. Irfan, and S. Samual, "8 Things Everyone Should Know About Australia's Wildlife Disaster," *Vox*, January 22, 2020.

42 "fire managers": Quote from David Bowman and Ross A. Bradstock, "Australia Needs a National Fire Inquiry—These Are the 3 Key Areas It Should Deliver In," *The Conversation*, January 22, 2020.

44 a scattering of people: Many sources examine the collapse of Mayan civilization, including B. L. Turner, "The Ancient Maya: Sustainability and Collapse?," in *Routledge Handbook of the History of Sustainability*, ed. Jeremy L. Caradonna (Abingdon, U.K.: Routledge, 2017), 57–68; and John Haldon et al., "History Meets Palaeoscience: Consilience and Collaboration in Studying Past Societal Responses to Environmental Change," *Proceedings of the National Academy of Sciences* 115, no. 13 (2018): 3210–3218.

44 too little food: For the many causal factors for collapse of civilizations and the pitfalls of single-causal explanations, see Guy D. Middleton, "Nothing Lasts Forever: Environmental Discourses on the Collapse of Past Societies," *Journal of Archaeological Research* 20, no. 3 (2012): 257–307; and Haldon et al., "History Meets Palaeoscience."

44 heat their homes: E. T. Wilkins, "Air Pollution and the London Fog of December, 1952," *Journal of the Royal Sanitary Institute* 74, no. 1 (1954): 1–21.

44 soot-laden skies: Spyros N. Pandis et al., "Urban Particulate Matter Pollution: A Tale of Five Cities," *Faraday Discussions* 189 (2016): 277–90, tells the stories of five once-polluted cities.

45 thick layers of smog: For a listing from World Health Organization on PM2.5 concentrations in cities in 2016, see www.who.int/airpollution/data/cities-2016/en. See Li Wang et al., "Taking Action on Air Pollution Control in the Beijing–Tianjin–Hebei (Bth) Region: Progress, Challenges and Opportunities," *International Journal of Environmental Research and Public Health* 15, no. 2 (2018): 306, for improvements of air quality in Beijing.

45 United Kingdom: Churchill quote from www.azquotes.com/quote/503675.

46 a global response: The World Health Organization's response plan is outlined in "Operational Framework for the Deployment of the World Health Organization Smallpox Vaccine Emergency Stockpile in Response to a Smallpox Event," (Geneva: World Health Organization, 2017).

46 "should emergency containment measures be needed": Quote from D. A. Henderson and Isao Arita, "The Smallpox Threat: A Time to Reconsider Global Policy," *Biosecurity and Bioterrorism: Biodefense Strategy, Practice, and Science* 12, no. 3 (2014): 117–21.

3. HEDGES FOR BETS: INVEST IN DIVERSITY

51 world's drylands: P. E. Rajasekharan, "Gene Banking for ex Situ Conservation of Plant Genetic Resources," in *Plant Biology and Biotechnology*, Vol. 2, *Plant Genomics and Biotechnology*, ed. Bir Bahadur, Manchikatla Venkat Rajam, Leela Sahijram, and K. V. Krishnamurthy (New Delhi: Springer,2015), 445–59.

52 "these resources": Jack Harlan's quote is from Jack R. Harlan, "Genetics of Disaster 1," *Journal of Environmental Quality* 1, no. 3 (1972): 212–15; and quoted in C. Fowler, *Seeds on Ice* (Westport, Conn.: Prospecta Press, 2016). Also see Rajasekharan, "Gene Banking for ex Situ Conservation of Plant Genetic Resources"; and Gea Galluzzi et al., "Twenty-Five Years of International Exchanges of Plant Genetic Resources Facilitated by the CGIAR Genebanks: A Case Study on Global Interdependence," *Biodiversity and Conservation* 25, no. 8 (2016): 1421–46. The international gene banks are maintained by the Consultative Group for International Agricultural Research, a global partnership to address food security. See www.cgiar.org.

52 Arctic Circle: The Aleppo gene bank is ICARDA (International Center for Agricultural Research in the Dry Areas), one of the CGIAR international gene banks. The rescue is described in L. Wade, "How Syrians Saved an Ancient Seedbank from Civil War," *Wired*, April 17, 2015.

53 lockdown mode: See Rajasekharan, "Gene Banking for ex Situ Conservation of Plant Genetic Resources"; Fowler, *Seeds on Ice*; and Pablo A. Pellegrini and Galo E. Balatti, "Noah's Arks in the XXI Century. A Typology of Seed Banks," *Biodiversity and Conservation* 25, no. 13 (2016): 2753–69.

53 "an insurance policy for the world": R. Schiffman, "The Seeds of the Future," *New Scientist* 225, no. 3002 (2015): 23.

53 "prepared for change": Fowler, *Seeds on Ice*, 153.

53 keep future prospects alive: J. Seabrook, "Sowing for Apocalypse: The Quest for a Global Seed Bank," *New Yorker*, August 20, 2007.

53 Terbol and Rabat: Schiffman, "The Seeds of the Future"; and Fowler, *Seeds on Ice*.

54 toothless flies to eat: M. Schapiro, "How Seeds from War-Torn Syria Could Help Save American Wheat," *YaleEnvironment360*, May 14, 2018.

54 lifeline for humanity: D. Carrington, "Arctic Stronghold of World's Seeds Flooded After Permafrost Melts," *The Guardian*, May 19, 2017.

54 distant lands: Pellegrini and Balatti, "Noah's Arks in the XXI Century." The Hortus Botanicus in Leiden in the Netherlands, the Jardin des Plantes in Paris, and the Royal Botanic Garden in Edinburgh date to the late sixteenth century.

54 "Soviet agriculture": Nikolaĭ Ivanovich Vavilov et al., *Origin and Geography of Cultivated Plants* (Cambridge: Cambridge University Press, 1992), xiv.

55 East Asiatic Center: G. L. Corinto, "Nikolai Vavilov's Centers of Origin of Cultivated Plants with a View to Conserving Agricultural Biodiversity," *Human Evolution* 29, no. 4 (2014): 285–301, provides crops in the centers of origin and other information about Vavilov. See also Colin K. Khoury et al., "Origins of Food Crops Connect Countries Worldwide," *Proceedings of the Royal Society of London B: Biological Sciences* 283, no. 1832 (2016): 20160792; and the chapter on "Phyto-geographical Basis for Plant Breeding," in Vavilov et al., *Origin and Geography of Cultivated Plants*, 316–66.

55 cultivated tomato: See José Esquinas-Alcázar, "Protecting Crop Genetic Diversity for Food Security: Political, Ethical and Technical

Challenges," *Nature Reviews Genetics* 6, no. 12 (2005): 946, box 1 for the tomato and other examples of improvements from crop wild relatives.

55 these achievements: Vavilov et al., *Origin and Geography of Cultivated Plants*, xx.

56 their salvation: Pellegrini and Balatti, "Noah's Arks in the XXI Century."

56 climate-controlled facilities: Schiffman, "The Seeds of the Future"; and H. F. Chin, "Don't Let It Go! Saving Endangered Plants Through Seed Storage," *UTAR Agriculture Science Journal* 2, no. 3 (2016): 57–60.

57 " when it began": Schiffman, "The Seeds of the Future," 23.

57 a half million species: The estimate of the number of species at 5 million plus or minus 3 million is from Mark J Costello, Robert M May, and Nigel E Stork, "Can We Name Earth's Species Before They Go Extinct?," *Science* 339, no. 6118 (2013): 413–16. They explain the refinements in previous estimates that were as high as 100 million. The estimate of one trillion species is from Kenneth J. Locey and Jay T Lennon. "Scaling Laws Predict Global Microbial Diversity," *Proceedings of the National Academy of Sciences* 113, no. 21 (2016): 5970–75. See Esquinas-Alcázar, "Protecting Crop Genetic Diversity for Food Security," for other wild tomato species that provided characteristics to improve cultivated tomatoes.

58 keep the planet habitable: Sean B. Carroll, "Chance and Necessity: The Evolution of Morphological Complexity and Diversity," *Nature* 409, no. 6823 (2001): 1102; Brendan B. Larsen et al., "Inordinate Fondness Multiplied and Redistributed: The Number of Species on Earth and the New Pie of Life," *Quarterly Review of Biology* 92, no. 3 (2017): 229–65; Camilo Mora et al., "How Many Species Are There on Earth and in the Ocean?," *PLoS Biology* 9, no. 8 (2011); and Stuart L. Pimm et al., "The Biodiversity of Species and Their Rates of Extinction, Distribution, and Protection," *Science* 344, no. 6187 (2014): e1001127.

58 from one set of dominating species to another: See Douglas Fox, "What Sparked the Cambrian Explosion?," *Nature News* 530, no. 7590 (2016): 268.

59 a more stable climate: Lenton, Watson, and Watson, *Revolutions That Made the Earth*; and Daniel S. Heckman et al., "Molecular Evidence for the Early Colonization of Land by Fungi and Plants," *Science* 293, no. 5532 (2001): 1129–33.

59 deep into the soil: C. Strullu-Derrien et al., "The Origin and Evolution of Mycorrhizal Symbiosis: From Palaeomycology to Phylogenomics," *New Phytologist* 220, no. 4 (2018): 1012–30.

60 "the same place": Quote from Lewis Carroll, *Through the Looking Glass: And What Alice Found There* (Chicago: Rand McNally, 1917), 39. The Red Queen hypothesis and coevolution is discussed in Richard Dawkins and John Richard Krebs, "Arms Races Between and Within Species," *Proceedings of the Royal Society of London B: Biological Sciences* 205, no. 1161 (1979): 489–511; Peter A. Abrams, "The Evolution of Predator-Prey Interactions: Theory and Evidence," *Annual Review of Ecology and Systematics* 31, no. 1 (2000): 79–105; Rachel M. Penczykowski, Anna-Liisa Laine, and Britt Koskella, "Understanding the Ecology and Evolution of Host–Parasite Interactions Across Scales," *Evolutionary Applications* 9, no. 1 (2016): 37–52; and Dieter Ebert, "Host–Parasite Coevolution: Insights from the Daphnia–Parasite Model System," *Current Opinion in Microbiology* 11, no. 3 (2008): 290–301. The Red Queen hypothesis was coined by Leigh Van Valen, "A New Evolutionary Law," *Evol Theory* 1 (1973): 1–30.

62 Maria-like hurricane: Colin M. Donihue et al., "Hurricane-induced Selection on the Morphology of an Island Lizard," *Nature* 560 (2018): 88–91. See a popular article about lizards and Hurricane Maria at E. Yong, "After Last Year's Hurricanes, Caribbean Lizards Are Better at Holding on for Dear Life," *Atlantic*, July 25, 2018.

63 individual investors: For more information about Harry Markowitz, see www.nobelprize.org/nobel_prizes/economic-sciences/laureates/1990/markowitz-facts.html.

63 "dissimilar industries": Harry Markowitz, "Portfolio Selection," *Journal of Finance* 7, no. 1 (1952): 77–91. A portfolio approach related to conservation and ecology is discussed in Daniel E. Schindler, Jonathan B. Armstrong, and Thomas E. Reed, "The Portfolio Concept in Ecology and Evolution," *Frontiers in Ecology and the Environment* 13, no. 5 (2015): 257–63.

64 "assumed to be true": John von Neumann, "Probabilistic Logics and the Synthesis of Reliable Organisms from Unreliable Components," *Automata Studies* 34 (1956): 43–98.

64 in case one fails: John Downer, *When Failure Is an Option: Redundancy, Reliability and Regulation in Complex Technical Systems* (London: Centre for Analysis of Risk and Regulation, London School of Economics and Political Science, 2009); Martin Landau, "Redundancy, Rationality, and the Problem of Duplication and Overlap," *Public Administration*

Review 29, no. 4 (1969): 346–58; Matthew J. Hawthorne and Dewayne E. Perry, "Applying Design Diversity to Aspects of System Architectures and Deployment Configurations to Enhance System Dependability" (paper presented at the DSN 2004 Workshop on Architecting Dependable Systems [DSN-WADS 2004], June 28–July 1, 2004 in Florence, Italy); and Bev Littlewood, Peter Popov, and Lorenzo Strigini, "Design Diversity: An Update from Research on Reliability Modelling," in *Aspects of Safety Management* , ed. Felix Redmill and Tom Anderson (London: Springer, 2001), 139–54.

65 volcanic ash: Downer, *When Failure Is an Option*, discusses redundancy in engineering systems and refers to the Jakarta incident. The incident is also the topic of many books and documentaries, including the book *All Four Engines Have Failed: The True and Triumphant Story of Flight BA 009 and the Jakarta Incident*, by Betty Tootell (Auckland: Hutchinson Group, 1985); a National Geographic show, http://www.nationalgeographic.com.au/tv/air-crash-investigation; and a magazine article, "When Volcanic Ash Stopped a Jumbo at 37.000ft," *BBC News Magazine*, April 15, 2010.

66 one alone: the number of species of butterflies and moths is derived from the pollination data set in Nina Joffard, François Massol, Matthias Grenié, Claudine Montgelard, and Bertrand Schatz, "Effect of Pollination Strategy, Phylogeny and Distribution on Pollination Niches of Euro-Mediterranean Orchids," *Journal of Ecology* 107, no. 1 (2019): 478–90. The nested assembly of animal-plant and organizational mutualistic networks is described in Jordi Bascompte, Pedro Jordano, Carlos J Melián, and Jens M Olesen, "The Nested Assembly of Plant–Animal Mutualistic Networks," *Proceedings of the National Academy of Sciences* 100, no. 16 (2003): 9383–87. Ecological and organizational networks are investigated in Serguei Saavedra, Felix Reed-Tsochas, and Brian Uzzi, "A Simple Model of Bipartite Cooperation for Ecological and Organizational Networks," *Nature* 457, no. 7228 (2009): 463–66.

67 the Gondi language: The number of Gondi speakers in the 2011 Census of India was 2,984,453; www.censusindia.gov.in/2011Census/Language-2011/Part-B.pdf. The total Gond population is 10,859,422 according to the 2001 census most recently available from www.tribal.nic.in/ST/Tribal Profile.pdf.

67 somewhere in the world: Estimates of the number of languages are provided in Ethnologue: Languages of the World, www.ethnologue.com; UNESCO Atlas of the World's Languages in Danger, www.unesco.org/languages-atlas; and Jonathan Loh and David Harmon, "Biocultural Diversity: Threatened Species, Endangered Languages" (Zeist, The Netherlands: WWF Netherlands, June 8, 2014).

68 plants and animals: Ethnologue, UNESCO Atlas, and Loh and Harmon, "Biocultural Diversity."

68 repositories of knowledge: The number of languages spoken in each country is provided in Ethnologue: Languages of the World.

68 Persian: Daniel Nettle, "Linguistic Diversity of the Americas Can Be Reconciled with a Recent Colonization," *Proceedings of the National Academy of Sciences* 96, no. 6 (1999): 3325–29; and Daniel Nettle, "Explaining Global Patterns of Language Diversity," *Journal of Anthropological Archaeology* 17, no. 4 (1998): 354–74.

68 Australia: Daniel Nettle and Suzanne Romaine, *Vanishing Voices: The Extinction of the World's Languages* (New York: Oxford University Press on Demand, 2000); and Nicholas Evans, *Dying Words: Endangered Languages and What They Have to Tell Us*, (Chichester, U.K.: Wiley-Blackwell, 2011).

69 July 22, 2016: Anon, "Last Speaker of Majhi Language Dead," *The Times of India*, July 22, 2016. Some Majhi speakers might still be alive in Nepal.

69 he wrote: Charles Darwin, *The Descent of Man and Selection in Relation to Sex*, vol. 1 (London: Murray, 1888), 126.

70 "killed and sold": Basil H Johnston, "One Generation from Extinction," *Native Writers and Canadian Writing* 1990: 10.

70 a clockwork view of the world: Johnstion, "One Generation," 12. See also Robin Kimmerer, *Braiding Sweetgrass: Indigenous Wisdom, Scientific Knowledge and the Teachings of Plants* (Minneapolis, Minn.: Milkweed Editions, 2013), for Native American traditions and experience with plants.

71 Kuuk Thaayorre language: Lera Boroditsky, "How Language Shapes Thought," *Scientific American* 304, no. 2 (2011): 64. Linguists debate about the degree to which differences in grammar structures influence thought processes. A view counter to the notion that language influences the way people view the world is provided in John H. McWhorter, *The Language Hoax: Why the World Looks the Same in Any Language* (New York: Oxford University Press, 2014).

71 Parkinson's disease: Many sources discuss the importance of traditional knowledge for drug discovery, among them are Graham Dutfield, "Opinion: Why Traditional Knowledge Is Important in Drug Discovery," *Future Medicinal Chemistry* 2, no. 9 (2010): 1405–409; T. de Sousa Araujo et al., "Medicinal Plants," in *Introduction to Ethnobiology*, ed. U. Albuquerque and R. Alves (Cham, Switzerland: Springer, 2016), 143–49; and Daniel S. Fabricant and Norman R. Farnsworth, "The Value of Plants Used in Traditional Medicine for Drug Discovery," *Environmental Health Perspectives* 109, no. S1 (2001): 69. The World Health Organization estimates that about 25 percent of all modern medicines are derived from medicinal plants. P. Stephens and H. Leufkens, "The World Medicines Situation 2011: Research and Development," ed. World Health Organization (Geneva World Health Organization, 2011), 1–28.

72 fractures in the ice: Fikret Berkes and Dyanna Jolly, "Adapting to Climate Change: Social-Ecological Resilience in a Canadian Western Arctic Community," *Conservation Ecology* 5, no. 2 (2002): art18; and Theresa Nichols et al., "Climate Change and Sea Ice: Local Observations from the Canadian Western Arctic," *Arctic* 57, no. 1 (2004): 68–79. The Inuvialuit are Inuit people who live in the western Canadian region.

72 one Inuvialuit: Berkes and Jolly, "Adapting to Climate Change," 7.

73 "hunger and deprivation": Quote from Norman Borlaug's acceptance speech "The Green Revolution, Peace, and Humanity" on being awarded the Nobel Peace Prize. Nobel Lecture, December 11, 1970, https://www.nobelprize.org/prizes/peace/1970/borlaug/lecture.

73 about 7,000: Food and Agriculture Organization, *The State of the World's Plant Genetic Resources for Food and Agriculture* (Rome: Food and Agriculture Organization, 1997).

74 "its culture": Jefferson quote from 1800 is recorded in the Summary of Public Service at founders.archives.gov/documents/Jefferson/01-32-02-0080. Other historical information in the paragraph is from Paul Gepts, "Plant Genetic Resources Conservation and Utilization," *Crop Science* 46, no. 5 (2006): 2278–92.

74 Thak Bahadu's last breath: The loss of landraces with the advent of Green Revolution varieties is discussed in Prabhu L. Pingali, "Green Revolution: Impacts, Limits, and the Path Ahead," *Proceedings of the National Academy of Sciences* 109, no. 31 (2012); and P. Pingali, "The

Green Revolution and Crop Diversity," in *Routledge Handbook of Agricultural Biodiversity*, ed. Danny Hunter et al. (Abingdon, U.K.: Routledge, 2017), chap. 12.

74 a few hundred remain today: Gepts, "Plant Genetic Resources Conservation and Utilization."

75 Indian village: Colin K. Khoury et al., "Increasing Homogeneity in Global Food Supplies and the Implications for Food Security," *Proceedings of the National Academy of Sciences* 111, no. 11 (2014): 4001–4006, analyze the homogenization of diets over time. Barry M. Popkin, "Global Nutrition Dynamics: The World Is Shifting Rapidly Toward a Diet Linked with Noncommunicable Diseases," *American Journal of Clinical Nutrition* 84, no. 2 (2006): 289–98, describes the nutritional transition producing similar trajectories in different countries and cultures.

76 path to destruction: Donald Craig Willcox, Giovanni Scapagnini, and Bradley J. Willcox, "Healthy Aging Diets Other Than the Mediterranean: A Focus on the Okinawan Diet," *Mechanisms of Ageing and Development* 136 (2014): 148–62; and D. Craig Willcox et al., "The Okinawan Diet: Health Implications of a Low-Calorie, Nutrient-Dense, Antioxidant-Rich Dietary Pattern Low in Glycemic Load," *Journal of the American College of Nutrition* 28, no. S4 (2009): 500S–516S.

76 "Takes a Big Dive": N. Onisha, "Love of U.S. Food Shortening Okinawans' Lives/Life Expectancy Among Islands' Young Men Takes a Big Dive," *New York Times*, April 4, 2004.

77 the human genome: L. M. Caetano Antunes et al., "The Effect of Antibiotic Treatment on the Intestinal Metabolome," *Antimicrobial Agents and Chemotherapy* 55, no. 4 (2011): 1494–1503; Elizabeth L. Johnson et al., "Microbiome and Metabolic Disease: Revisiting the Bacterial Phylum Bacteroidetes," *Journal of Molecular Medicine* 95, no. 1 (2017): 1–8; Muhammad Shahid Riaz Rajoka et al., "Interaction Between Diet Composition and Gut Microbiota and Its Impact on Gastrointestinal Tract Health," *Food Science and Human Wellness* 6, no. 3 (2017): 121–30; and Maryam Tidjani Alou, Jean-Christophe Lagier, and Didier Raoult, "Diet Influence on the Gut Microbiota and Dysbiosis Related to Nutritional Disorders," *Human Microbiome Journal* 1 (2016): 3–11.

77 our microbiomes: Francisco Guarner, "Decade in Review—Gut Microbiota: The Gut Microbiota Era Marches On," *Nature Reviews Gastroenterology and Hepatology* 11, no. 11 (2014): 647–49.

77 able to survive: Quoted in Guarner, "Decade in Review," from Louis Pasteur, "Observations Relative a La Note De M. Duclaux," *Comptes rendus de l'Académie des Sciences* 100 (1885): 68.

78 "the gut microbiome": Quote from an interview with Jeffrey Gordon at www.pmwcintl.com/jeffrey-gordon-qa.

78 the American gut: Tanya Yatsunenko et al., "Human Gut Microbiome Viewed across Age and Geography," *Nature* 486, no. 7402 (2012): 222.

78 obesity, digestive diseases, and cancer: Ruth E. Ley et al., "Microbial Ecology: Human Gut Microbes Associated with Obesity," *Nature* 444, no. 7122 (2006): 1022; Mark L. Heiman and Frank L. Greenway, "A Healthy Gastrointestinal Microbiome Is Dependent on Dietary Diversity," *Molecular Metabolism* 5, no. 5 (2016): 317–20; and Alexis Mosca, Marion Leclerc, and Jean P. Hugot, "Gut Microbiota Diversity and Human Diseases: Should We Reintroduce Key Predators in Our Ecosystem?," *Frontiers in Microbiology* 7 (2016): 455.

79 keep people healthy: Heiman and Greenway, "A Healthy Gastrointestinal Microbiome Is Dependent on Dietary Diversity."

79 spirochete bacteria: The history of Paul Ehrlich's discovery of a cure for syphilis is discussed in A. Lennox Thorburn, "Paul Ehrlich: Pioneer of Chemotherapy and Cure by Arsenic (1854–1915)," *Sexually Transmitted Infections* 59, no. 6 (1983): 404–405; D. M. Jolliffe, "A History of the Use of Arsenicals in Man," *Journal of the Royal Society of Medicine* 86, no. 5 (1993): 287; and Rustam I. Aminov, "A Brief History of the Antibiotic Era: Lessons Learned and Challenges for the Future," *Frontiers in Microbiology* 1 (2010): 134.

80 "my culture plate": Siang Yong Tan and Yvonne Tatsumura, "Alexander Fleming (1881–1955): Discoverer of Penicillin," *Singapore Medical Journal* 56, no. 7 (2015): 366, discuss the history of Alexander Fleming's discovery of penicillin. Quote from Sir Alexander Flemings's banquet speech on his acceptance of the Nobel Prize on December 10, 1945. The full speech is available at www.nobelprize.org/prizes/medicine/1945/fleming/speech.

80 the present day: Aminov, "A Brief History of the Antibiotic Era"; Martin J. Blaser, "Antibiotic Use and Its Consequences for the Normal Microbiome," *Science* 352, no. 6285 (2016): 544–45; Morten O. A. Sommer and Gautam Dantas, "Antibiotics and the Resistant Microbiome," *Current Opinion in Microbiology* 14, no. 5 (2011): 556–63; Antunes et al., "The Effect of Antibiotic Treatment on the Intestinal Metabolome";

and Andreas J. Bäumler and Vanessa Sperandio, "Interactions Between the Microbiota and Pathogenic Bacteria in the Gut," *Nature* 535, no. 7610 (2016): 85.

4. MIND THE NET: DEFEND AGAINST CASCADING FAILURE

84 storage facilities: Daniel M. Bernhofen, Zouheir El-Sahli, and Richard Kneller, "Estimating the Effects of the Container Revolution on World Trade," *Journal of International Economics* 98 (2016): 36–50.

85 "who changed the world": B. Rascovar, "Shipping Pioneer Largely Ignored," *The Sun*, June 14, 2001.

85 a global economy: Christopher Chase-Dunn, Yukio Kawano, and Benjamin D. Brewer, "Trade Globalization Since 1795: Waves of Integration in the World-System," *American Sociological Review* 65, no. 1 (2000): 77–95.

86 food distribution network: New York City Economic Development Council and Mayor's Office of Recovery and Resiliency, "Five Borough Food Flow: 2016 New York City Food Distribution and Resiliency Study Results" (New York: New York City Economic Development Council and Mayor's Office of Recovery and Resiliency, 2016).

86 mountain paths: Paul Beynon-Davies, "Informatics and the Inca," *International Journal of Information Management* 27, no. 5 (2007): 306–18.

87 a nationwide blackout: S. Karanja, "Monkey Tripped Transformer at Power Station Causing Massive Blackout, Kengen Says," *Daily Nation*, June 8, 2016.

87 three hours: J. Gettleman, "Monkey in Kenya Survives After Setting Off Nationwide Blackout," *New York Times*, June 8, 2016.

87 safeguards were in place: Harvey J. Alter and Harvey G. Klein, "The Hazards of Blood Transfusion in Historical Perspective," *Blood* 112, no. 7 (2008): 2617–26; Elaine M. Sloand, Elisabeth Pitt, and Harvey G. Klein, "Safety of the Blood Supply," *JAMA* 274, no. 17 (1995): 1368–73; and Miguel A. Centeno et al., "The Emergence of Global Systemic Risk," *Annual Review of Sociology* 41 (2015): 65–85.

87 computers around the world: Steven H. Strogatz, "Exploring Complex Networks," *Nature* 410, no. 6825 (2001): 268–76. Duncan J. Watts, "A Simple Model of Global Cascades on Random Networks," *Proceedings*

of the National Academy of Sciences 99, no. 9 (2002): 5766–71, analyzes theoretical aspects of cascading failures in networks.

89 nodes and thread as links: Stuart Kauffman, "At Home in the Universe: The Search for the Laws of Self-Organization and Complexity" (New York: Oxford University Press 1996), 320. The button metaphor is also explained in Strogatz, "Exploring Complex Networks." The science of networks dates back to the Swiss logician Leonhard Euler, who in 1736 used a wiring diagram to analyze bridges in the city of Königsberg, the capital of Eastern Prussia.

89 social networks: Six degrees of separation is the popular notion that people anywhere in the world are connected to each other through six or fewer steps. Stanley Milgram carried out an experiment in 1967 based on earlier work to test the number of steps required to connect people through social networks. Milgram randomly selected people in the Midwest and asked them to send a package to a stranger in Massachusetts by sending it to someone they knew on a first-name basis who might be connected to the target person. Milgram reported that on average it took between five and seven intermediaries to reach the target. Jeffrey Travers and Stanley Milgram, "The Small World Problem," *Psychology Today* 1 (1967): 61–67. The findings were later disputed as an "urban myth." Judith S. Kleinfeld, "Six Degrees of Separation: Urban Myth?," *Psychology Today* 35, no. 2 (2002): 74. The application of the concept to other types of networks is explored in Duncan Watts, *Six Degrees: The Science of a Connected Age* (New York: Norton, 2003), 376. The term "small world" is often used loosely. In this chapter, it means a network structure that has a high clustering of nodes and a small number of nodes with a high degree (i.e., a large number of links).

90 small-world pattern: Duncan J. Watts and Steven H. Strogatz, "Collective Dynamics of 'Small-World' Networks," *Nature* 393, no. 6684 (1998): 440–42. In this highly influential paper, the authors report results of generating a small-world network by "rewiring" a small number of links in a random network.

90 a small number of partners: Fredrik Liljeros et al., "The Web of Human Sexual Contacts," *Nature* 411, no. 6840 (2001): 907–908; and James Holland Jones and Mark S. Handcock, "An Assessment of Preferential Attachment as a Mechanism for Human Sexual Network Formation," *Proceedings of the Royal Society of London B: Biological Sciences* 270, no. 1520 (2003): 1123–28.

90 another small airport: Albert-László Barabási and Réka Albert, "Emergence of Scaling in Random Networks," *Science* 286, no. 5439 (1999): 509–12. In this paper, the authors describe the emergence of scale-free networks. This paper is one of the most highly cited papers in network science.

91 "100 feet tall": Albert-László Barabási and Eric Bonabeau, "Scale-Free," *Scientific American* 288, no. 5 (2003): 50–59. This paper on scale-free networks is very accessible to the lay reader.

91 in the extreme: A network with a few hubs with extremely large numbers of connections and the vast majority with very few connections is referred to as "scale-free." The term reflects the freedom from a typical node, unlike a typical height in a bell-shaped curve. The distribution of degrees (number of links for each node) follows a power-law distribution in a scale-free network. Consequently, there is no meaningful average value as there is in a Poisson distribution of random networks. In power-law distribution, a small number of nodes have a very large number of links. Power-law distributions have been observed in income distributions, earthquake magnitudes, word frequencies, and many other phenomena.

91 financial markets: There is controversy whether power grid networks are scale-free. Giuliano Andrea Pagani and Marco Aiello, "Power Grid Complex Network Evolutions for the Smart Grid," *Physica A: Statistical Mechanics and Its Applications* 396 (2014): 248–66.

92 the whole network still functions: Vito Latora and Massimo Marchiori, "Efficient Behavior of Small-World Networks," *Physical Review Letters* 87, no. 19 (2001): 198701, discuss the efficiency of small-world networks. These networks are efficient yet vulnerable to attack on a hub.

92 proved costly: T. Fuller, "Thailand Flooding Crippled Hard-Drive Suppliers," *New York Times*, November 6, 2011.

93 life in isolation: Filio Marineli et al., "Mary Mallon (1869–1938) and the History of Typhoid Fever," *Annals of Gastroenterology* 26, no. 2 (2013): 132.

93 before finally subsiding: Zhuang Shen et al., "Superspreading Sars Events, Beijing, 2003," *Emerging Infectious Diseases* 10, no. 2 (2004): 256; Lee Shiu Hung, "The SARS Epidemic in Hong Kong: What Lessons Have We Learned?," *Journal of the Royal Society of Medicine* 96, no. 8 (2003): 374–78; and World Health Organization, "SARS: Lessons from a New Disease," in *The World Health Report 2003—Shaping the Future* (Geneva, Switzerland: WHO, 2005), 73–82. Another super-spreader

for HIV AIDS was a gay Canadian flight attendant who had sex with thousands of men. He was vilified as patient zero for spreading AIDS in New York in the 1980s, but later vindicated when genetic tracers of the virus showed that he was not patient zero. Michael Worobey et al., "1970s and 'Patient 0' HIV-1 Genomes Illuminate Early HIV/AIDS History in North America," *Nature* 539, no. 7627 (2016): 98–101.

94 longer than humans: Heckman et al., "Molecular Evidence for the Early Colonization of Land by Fungi and Plants." *Science* 293, no. 5532 (2001): 1129–33.

94 from Seoul to New York: P. Crane, *Gingko: The Tree That Time Forgot* (New Haven, Conn.: Yale University Press, 2013).

95 thicker, primary ones: Anita Roth-Nebelsick et al., "Evolution and Function of Leaf Venation Architecture: A Review," *Annals of Botany* 87, no. 5 (2001): 553–66; Lawren Sack and Christine Scoffoni, "Leaf Venation: Structure, Function, Development, Evolution, Ecology and Applications in the Past, Present and Future," *New Phytologist* 198, no. 4 (2013): 983–1000; and Lawren Sack et al., "Leaf Palmate Venation and Vascular Redundancy Confer Tolerance of Hydraulic Disruption," *Proceedings of the National Academy of Sciences* 105, no. 5 (2008): 1567–72.

96 a branching tree: Eleni Katifori, Gergely J Szöllősi, and Marcelo O. Magnasco, "Damage and Fluctuations Induce Loops in Optimal Transport Networks," *Physical Review Letters* 104, no. 4 (2010): 048704. Also see fascinating videos of movement of dye through leaf veins in a YouTube video titled *Lighting up Leaves*, https://www.youtube.com/watch?v=Gps8uktGtFY; and in "Why Leaves Aren't Trees," *Physics Review Focus*, https://physics.aps.org/story/v25/st4.

96 too much for the veins to handle: Francis Corson, "Fluctuations and Redundancy in Optimal Transport Networks," *Physical Review Letters* 104, no. 4 (2010): 048703.

97 keep the wing strong: Eleni Katifori and Marcelo O. Magnasco, "Quantifying Loopy Network Architectures," *PLoS ONE* 7, no. 6 (2012): e37994.

98 efficiency and fragility: Charles A. Price and Joshua S. Weitz, "Costs and Benefits of Reticulate Leaf Venation," *BMC Plant Biology* 14, no. 1 (2014): 234.

98 command and control center: In a 2001 interview in *Wired* magazine, Paul Baran says that the notion that Arpanet (the precursor to the

internet) was developed to withstand nuclear strikes is a myth. Rather, it was developed to connect computer terminals so they could talk to each other. See S. Brand, "Founding Father: *Wired* Legends," *Wired*, March 1, 2001, for the interview.

98 "hot-potato routing": Brand, "Founding Father."

100 Baran's foresight: Paul Baran's reports are available from the Rand Corporation at www.rand.org/pubs/research_memoranda/RM3420.html.

100 "simply unheard of": James Pelkey, "Entrepreneurial Capitalism and Innovation: A History of Computer Communications 1968–1988" (2007), www. historyofcomputercommunications; quote is from sect. 4.12. For the history of the internet, see Barry M. Leiner et al., "A Brief History of the Internet," *ACM SIGCOMM Computer Communication Review* 39, no. 5 (2009): 22–31.

100 "in the 1960s": Quote from interview with Vint Cerf in R. Singel, "Vint Cerf: We Knew What We Were Unleashing on the World," *Wired*, April 23, 2012.

100 in 1964: Paul Baran, "On Distributed Communications: I. Introduction to Distributed Communicaitons Networks," (Santa Monica, CA: Rand Coporation Memorandum RM-3420-PR, August 1964), 16.

101 October 1789: Louise A. Tilly, "The Food Riot as a Form of Political Conflict in France," *Journal of Interdisciplinary History* 2, no. 1 (1971): 23–57; and J. Bohstedt, "Food Riots and the Politics of Provisions in World History" (Brighton, U.K.: Institute of Development Studies Working Paper No. 444, 2014).

101 labor in factories: John Bohstedt, *The Politics of Provisions: Food Riots, Moral Economy, and Market Transition in England, c. 1550–1850* (Farnham, U.K.: Ashgate, 2013).

101 starved millions: Bohstedt, "Food Riots and the Politics of Provisions in World History."

102 "some developing countries": Quote from Mark Ritchie, *Impact of Gatt on World Hunger* (Minneapolis: Institute for Agriculture and Trade Policy, 1988), from April 18, 1983, interview with Secretary of Agriculture John Block reported in Capital Press. Several sources, cite a similar quote from John Block at the opening of the 1986 GATT meeting in Punta del Este, Uruguay, but his speech cannot be verified in the extensive GATT documentation (www.wto.org/english/docs_e/gattdocs_e.htm). See K. Malhotra, "The Uruguay Round of Gatt, the

World Trade Organization and Small Farmers" (paper prepared for the Regional Conference on MonoCultural Cropping in Southeast Asia: Social/Environmental Impacts and Sustainable Alternatives, Songkhla, Thailand, June 3–6, 1996); E. Goldsmith et al., "Cakes and Caviar? The Dunkel Draft and Third World Agriculture," *Ecologist* 23, no. 6 (1993): 219–22; and Raj Patel and Philip McMichael, "A Political Economy of the Food Riot," *Review: A Journal of the Fernand Braudel Center* 32, no. 1 (2009): 9–35.

102 "poor crops in others": Quote from Multilateral Trade Negotiations: The Uruguay Round, "Elaboration of Us Agricultural Proposal with Respect to Food Security Submitted by the United States, Negotiating Group on Agriculture, Group of Negotiations on Goods (Gatt), Mtn. Gng/Ng5/W/61, 6 June 1988" (1988). For the history of negotiations on agriculture in the General Agreement on Trade and Tariffs (GATT), which later became the World Trade Organization, see Zuhair A. Hassan, "Agreement on Agriculture in the Uruguay Round of GATT: From Punta Del Este to Marrakesh," *Agricultural Economics* 15, no. no. 1 (1996): 29–46; Anthony J. Rayner, K. A. Ingersent, and R. C. Hine, "Agriculture in the Uruguay Round: An Assessment," *Economic Journal* 103, no. 421 (1993): 1513–27; Jane M. Porter and Douglas E. Bowers, "A Short History of US Agricultural Trade Negotiations" (Washington, D.C.: U.S. Department of Agriculture–Economic Research Service, 1989); and Congressional Budget Office, "The Gatt Negotiations and U.S. Trade Policy" (Washington, D.C.: Congress of the United States, 1987).

102 in the coalition: Rod Tyers, "The Cairns Group and the Uruguay Round of International Trade Negotiations," *Australian Economic Review* 26, no. 1 (1993): 49–60.

102 since World War II: quotes from Robert J. McCartney, "Mexico to Lower Trade Barriers, Join Gatt," *Washington Post*, November 26, 1985.

103 tortillas skyrocket: Alder Keleman and Hugo García Rañó, "The Mexican Tortilla Crisis of 2007: The Impacts of Grain-Price Increases on Food-Production Chains," *Development in Practice* 21, no. 4–5 (2011): 550–65; and Lutz Kilian, "Explaining Fluctuations in Gasoline Prices: A Joint Model of the Global Crude Oil Market and the Us Retail Gasoline Market," *The Energy Journal* 31, no. 2 (2010) 87–112.

103 February 1, 2007: Elisabeth Malkin, "Thousands in Mexico City Protest Rising Food Prices," *New York Times*, February 1, 2007.

104 soaring food prices: L. Doyle, "Starving Haitians Riot as Food Prices Soar," *The Independent*, April 9, 2008; and Walden Bello and Mara Baviera, "Food Wars," *Monthly Review* 61, no. 3 (2009).

104 technological advances on farms: See Ruth DeFries, *The Big Ratchet: How Humanity Thrives in the Face of Natural Crisis* (New York: Basic Books, 2014), for the history of the advances in food production that led to falling food prices.

104 purchase food: See Derek Headey and Shenggen Fan, "Anatomy of a Crisis: The Causes and Consequences of Surging Food Prices," *Agricultural Economics* 39, no. S1 (2008): 375–91, for an overview of trends in food prices.

105 twenty years earlier: Clinton quote from "Bill Clinton: 'We Blew It' on Global Food," *CBS News*, October 23, 2008. Sources for responses during the 2008 crisis are Antoine Bouët and David Laborde Debucquet, "Food Crisis and Export Taxation: Revisiting the Adverse Effects of Noncooperative Aspect of Trade Policies," in *Food Price Volatility and Its Implications for Food Security and Policy*, ed. Matthias Kalkuhl, Joachim von Braun, and Maximo Torero, 167–79 (New York: Springer, 2016); Steve Wiggins and Sharada Keats, "Looking Back, Peering Forward: Food Prices and the Food Price Spike of 2007/08," (London: Overseas Development Institute, March 28, 2013), www. odi. org/sites/odi. org. uk/files/odi-assets/publications-opinion-files/8339. pdf 2013; and Will Martin and Kym Anderson, *Export Restrictions and Price Insulation During Commodity Price Booms* (Washington, D.C.: World Bank, 2011).

105 the world stage: Caitlin E. Werrell, Francesco Femia, and Troy Sternberg, "Did We See It Coming? State Fragility, Climate Vulnerability, and the Uprisings in Syria and Egypt," *SAIS Review of International Affairs* 35, no. 1 (2015): 29–46.

105 a long, rocky path: Maximo Torero, "Alternative Mechanisms to Reduce Food Price Volatility and Price Spikes: Policy Responses at the Global Level," in *Food Price Volatility and Its Implications for Food Security and Policy*, ed. Matthias Kalkuhl, Joachim von Braun, and Maximo Torero, 115–38. Cham, Switzerland: Springer, 2016), discusses the multiple efforts and proposals to reduce volatility in global food trade.

106 115 countries: Data on grain exports is for 2013 (most recent available) from FAOSTAT, www.fao.org/faostat/en, accessed July 3, 2017. See also Michael J. Puma et al., "Assessing the Evolving Fragility of the Global

Food System," *Environmental Research Letters* 10, no. 2 (2015): 024007; and Mária Ercsey-Ravasz et al., "Complexity of the International Agro-Food Trade Network and Its Impact on Food Safety," *PLoS ONE* 7, no. 5 (2012): e37810.

106 global food system: Dependence on trade has doubled since the 1980s. Philippe Marchand et al., "Reserves and Trade Jointly Determine Exposure to Food Supply Shocks," *Environmental Research Letters* 11, no. 9 (2016): 095009. Franziska Gaupp et al., "Dependency of Crop Production Between Global Breadbaskets: A Copula Approach for the Assessment of Global and Regional Risk Pools," *Risk Analysis* 37, no. 11 (2017): 2212–28, examine the role of crop insurance to alleviate risks in the global trade network for wheat. Jessica A. Gephart et al., "Vulnerability to Shocks in the Global Seafood Trade Network," *Environmental Research Letters* 11, no. 3 (2016): 035008, address the vulnerability of the global seafood trade. Also see the assessment of growth associated with efficient and redundant networks in global food trade by Ali Kharrazi, Elena Rovenskaya, and Brian D. Fath, "Network Structure Impacts Global Commodity Trade Growth and Resilience," *PLoS ONE* 12, no. 2 (2017): e0171184; and the examinations of vulnerability in the global trade network for mineral resources by Peter Klimek, Michael Obersteiner, and Stefan Thurner, "Systemic Trade Risk of Critical Resources," *Science Advances* 1, no. 10 (2015): e1500522.

106 send prices soaring: Laura Wellesley et al., "Chokepoints in Global Food Trade: Assessing the Risk," *Research in Transportation Business & Management* 25 (2017): 15–28. Singapore, for example, imports 70 percent of its maize, rice, soybean, and wheat, much of which passes through the Strait of Malacca.

107 locally produced: Stephen A. Wood et al., "Trade and the Equitability of Global Food Nutrient Distribution," *Nature Sustainability* 1, no. 1 (2018): 34.

108 the entire colony: S. Cremer, "Social Immunity in Insects," *Current Biology* 29, no. 11 (2019): R425–R473; James F. A. Traniello, Rebeca B. Rosengaus, and Keely Savoie, "The Development of Immunity in a Social Insect: Evidence for the Group Facilitation of Disease Resistance," *Proceedings of the National Academy of Sciences* 99, no. 10 (2002): 6838–42; and Rebeca B. Rosengaus et al., "Immunity in a Social Insect," *Naturwissenschaften* 86, no. 12 (1999): 588–91.

110 death of the colony: The strategies of social insects to fight the spread of pathogens are discussed in L. Liu et al., "The Mechanisms of Social Immunity Against Fungal Infections of Eusocial Insects," *Toxins* 11, no. 244 (2019): 1–21; Nathalie Stroeymeyt, Barbara Casillas-Pérez, and Sylvia Cremer, "Organisational Immunity in Social Insects," *Current Opinion in Insect Science* 5 (2014): 1–15; Michael Simone-Finstrom, "Social Immunity and the Superorganism: Behavioral Defenses Protecting Honey Bee Colonies from Pathogens and Parasites," *Bee World* 94, no. 1 (2017): 21–29; and Sylvia Cremer and Michael Sixt, "Analogies in the Evolution of Individual and Social Immunity," *Philosophical Transactions of the Royal Society of London B: Biological Sciences* 364, no. 1513 (2008): 129–42.

110 measles: Nathan D. Wolfe, Claire Panosian Dunavan, and Jared Diamond, "Origins of Major Human Infectious Diseases," *Nature* 447, no. 7142 (2007): 279.

110 "in its wake": World Health Assembly, "Thirty-Third World Health Assembly, Geneva 5–23 May 1980: Resolutions and Decisions, Annexes" (Geneva: World Health Organization, 1980).

110 killed many millions: Egyptian prisoners passed the virus to the Hittites during the Egyptian Hittite wars around 1350 BC. Athens suffered an epidemic in 430 BC. Alexander the Great's army was hit with the virus while invading the Indian subcontinent in 327 BC. Around AD 180, a large-scale epidemic ravaged the Roman Empire in the first stages of the great civilization's decline. Brian J. Simmons et al., "Smallpox: 12 000 Years from Plagues to Eradication: A Dermatologic Ailment Shaping the Face of Society," *JAMA Dermatology* 151, no. 5 (2015): 521.

111 early sixteenth century: Sergei N. Shchelkunov, "Emergence and Reemergence of Smallpox: The Need for Development of a New Generation Smallpox Vaccine," *Vaccine* 29 (2011): D49–D53. In an early act of biological warfare, the commander-in-chief of British forces in North America, Sir Jeffrey Amherst, suggested that the British grind scabs of smallpox pustules into blankets and distribute them to troublesome natives. Nicolau Barquet and Pere Domingo, "Smallpox: The Triumph Over the Most Terrible of the Ministers of Death," *Annals of Internal Medicine* 127, no. 8, part 1 (1997): 635–42; Frank Fenner et al., *Smallpox and Its Eradication* (Geneva: World Health Organization, 1988); and Donald A. Henderson, "The Eradication of Smallpox," *Scientific American* 235, no. 4 (1976): 25–33.

111 lucky enough to survive: Ralph W. Nicholas, "The Goddess Śītalā and Epidemic Smallpox in Bengal," *Journal of Asian Studies* 41, no. 1 (1981): 21–44. In the Middle Ages, the first medical description of smallpox, *De Variolis et Mobillis*, by the Persian physician Muhammad ibn Zakariya al-Razi, made two astute and pivotal observations. Those who survived the disease had lifelong immunity, and the disease was only transmitted person to person rather than from animal to human. Barquet and Domingo, "Smallpox: The Triumph over the Most Terrible of the Ministers of Death."

111 trigger an epidemic: Michael Radetsky, "Smallpox: A History of Its Rise and Fall," *Pediatric Infectious Disease Journal* 18, no. 2 (1999): 85–93. Lady Mary Montagu, a popular London aristocrat, inoculated her six-year-old son and three-year-old daughter during an epidemic, breaking the stigma of folk practices among the royals and wealthy of Europe and paving the way for vaccines.

111 eighteenth-century England: Barquet and Domingo, "Smallpox."

112 the Americas: For a history of Jenner's discovery and the acceptance of inoculation, see Barquet and Domingo, "Smallpox"; Daniel DiMaio, "Thank You, Edward. Merci, Louis," *PLoS Pathogens* 12, no. 1 (2016): e1005320; Fenner et al., *Smallpox and Its Eradication*; Henderson, "The Eradication of Smallpox"; Donald A. Henderson, "Principles and Lessons from the Smallpox Eradication Programme," *Bulletin of the World Health Organization* 65, no. 4 (1987): 535; and Radetsky, "Smallpox."

112 the other shore: Barquet and Domingo, "Smallpox." When King Charles IV of Spain decided to send the smallpox vaccine to his subjects in the Americas and Asia, the question was how to transport it across the ocean. The solution was to use orphans as human carriers.

112 each inoculation: Fenner et al., *Smallpox and Its Eradication*.

112 the deadly disease was in sight: For Louis Pasteur's accidental discovery of attenuated vaccines to replace vaccines taken from live virus and bacteria, see P. Berche, "Louis Pasteur, from Crystals of Life to Vaccination," *Clinical Microbiology and Infection* 18 (2012): 1–6; and Stanley A. Plotkin, "Vaccines: Past, Present and Future," *Nature Medicine* 11, no. S4 (2005): S5.

113 the disease still occurred: Frank Fenner, "Global Eradication of Smallpox," *Reviews of infectious diseases* 4, no. 5 (1982): 916–30.

113 ambitious endeavor: L. H. Collier developed freeze-dried vaccines. Fenner et al., *Smallpox and Its Eradication*.

113 "the rest of the population": Henderson, "The Eradication of Smallpox.", p. 30-31

114 contact with the carrier: Henderson, "The Eradication of Smallpox," describes the final days of the eradication campaign.

114 in 2013: A. Deria et al., "The World's Last Endemic Case of Smallpox: Surveillance and Containment Measures," *Bulletin of the World Health Organization* 58, no. 2 (1980): 279; and Michaeleen Doucleff, "Last Person To Get Smallpox Dedicated His Life To Ending Polio," *NPR Shots*, July 31, 2013, www.npr.org/sections/health-shots/2013/07/31/206947581/last-person-to-get-smallpox-dedicated-his-life-to-ending-polio.

115 "excessive optimism": Fenner et al., *Smallpox and Its Eradication*, 373.

115 "a university ivory tower": D. A. Henderson and Petra Klepac, "Lessons from the Eradication of Smallpox: An Interview with DA Henderson," *Philosophical Transactions of the Royal Society of London B: Biological Sciences* 368, no. 1623 (2013): 6.

115 eradicate diseases: Donald A. Henderson, "Lessons from the Eradication Campaigns," *Vaccine* 17 (1999): S53–S55.

115 the program failed badly: Fenner et al., "Smallpox and Its Eradication"; and Peter J. Hotez et al., "Hookworm: 'The Great Infection of Mankind.' " *PLoS Medicine* 2, no. 3 (2005): e67.

115 eradicate the disease: The five-year global campaign focused on cities and large towns on the premise that only a large human population could maintain an epidemic. Yellow fever ceased in cities in Ecuador within six months, Peru in 1921, and Central America in 1924. Health officials reported no new cases in all of the Americas in April 1927. In March 1928, after twenty years of no cases in Brazil, an outbreak took hold in the northeastern part of the country, then in the capital Rio de Janeiro, and in Colombia and Venezuela Fenner et al., *Smallpox and Its Eradication*.

115 "eradication of malaria": World Health Assembly, "Eighth World Health Assembly: Mexico D.F., 10–27 May 1955," (Geneva: World Health Organization, 1955), 31. The goal seemed possible. In Natal, Brazil—the unfortunate town where malaria-carrying mosquitoes had traveled through the mail from Dakar, Senegal, in the 1930s—a building-to-building campaign to rout out and douse mosquito breeding sites with insecticides had wiped out the mosquitoes in two years. Egypt had similar success in the mid-1940s with the same approach. Fenner et al., *Smallpox and Its Eradication*.

116 debilitating diseases: Malaria was gone from the United States, Greece, Italy, and Sardinia. Deaths from malaria plummeted in Sri Lanka, then Ceylon, Venezuela, and countries around the world. Doubts started to creep in not long after the program began. Successes occurred only in countries in developed parts of the world or where mosquitoes did not breed year-round. African countries south of the Sahara, a large part of the world's population that suffers from the most deadly form of malaria, had not been in the program from the beginning. Like yellow fever, the parasite could hide in monkeys and come back to infect humans. Most damaging, the parasite evolved to be resistant to DDT and drugs to treat the disease. Political support wavered with rising costs and less to show for the effort. In 1969, the World Health Assembly halted the coordinated global effort. Malaria rebounded in many countries as attention to control its spread slipped Fenner et al., *Smallpox and Its Eradication.*

116 this writing: Michael J. Toole, "So Close: Remaining Challenges to Eradicating Polio," *BMC Medicine* 14, no. 1 (2016): 43. The profile for polio fits for an eradicable disease. No animal reservoir harbors the virus and health workers can easily deliver an effective vaccine. Since 1988, when the World Health Assembly declared the initiative to eradicate polio, millions of health workers have fanned out to track down polio cases in remote places where health services seldom reach. Success was rapid. The last case in the Americas occurred in 1993, in the western Pacific in 1997, and in Africa in 2014. The virus stubbornly remains circulating in two countries, Pakistan and Afghanistan, proving the point that the last-mile obstacles are human as much as biological ones, in the form of local people's distrust of health workers and fading global commitments.

116 burrow into skin: David Molyneux and Dieudonné P. Sankara, "Guinea Worm Eradication: Progress and Challenges—Should We Beware of the Dog?," *PLoS Neglected Tropical Diseases* 11, no. 4 (2017): e0005495.

116 measles: Alan R. Hinman, "Measles and Rubella Eradication," *Vaccine* 36, no. 1 (2018): 1–3.

116 coming decades: After a forty-year hiatus, the quest to eradicate malaria is back on the table. The new push was boosted by success through drained swamps, screens on windows and doors, and house-to-house spraying to eliminate malaria from one hundred countries, including

the United States, over the last century. The success in some parts of the world was coupled with failure to make progress toward eliminating malaria in poorer areas. Gretchen Newby et al., "The Path to Eradication: A Progress Report on the Malaria-Eliminating Countries," *Lancet* 387, no. 10029 (2016): 1775–84. In 2007, the world's most influential living philanthropists, Bill and Melinda Gates, announced an "audacious goal." Melinda Gates called on world leaders to "reach a day when no human being has malaria, and no mosquito on earth is carrying it." www.gatesfoundation.org/media-center/speeches/2007/10/melinda-french-gates-malaria-forum.

116 1940 and 2005: Kate E. Jones et al., "Global Trends in Emerging Infectious Diseases," *Nature* 451, no. 7181 (2008): 990.

116 in 1999: Caren Chancey et al., "The Global Ecology and Epidemiology of West Nile Virus," *BioMed Research International* 2015 (2015): ar376230. https://doi.org/10.1155/2015/376230.

116 pathogen-carrying ticks: A. Marm Kilpatrick et al., "Lyme Disease Ecology in a Changing World: Consensus, Uncertainty and Critical Gaps for Improving Control," *Philosophical Transactions of the Royal Society of London B: Biological Sciences* 372, no. 1722 (2017): 20160117.

116 in 2015: Isaac I. Bogoch et al., "Anticipating the International Spread of Zika Virus from Brazil," *Lancet* 387, no. 10016 (2016): 335–36.

116 on the rise: Toph Allen et al., "Global Hotspots and Correlates of Emerging Zoonotic Diseases," *Nature Communications* 8, no. 1 (2017): 1124.

117 modular networks: B. Cowling and W. W. Lim, "They've Containted the Coronavirus. Here's How," *New York Times*, March 13, 2020.

117 neighboring towns: The story of Gunnison, Colorado, is described in Anon, "Gunnison and the Great Influenza," *Gunnison Country Times* (Gunnison, Colo.), January 18, 2018. H. Markel et al., "Nonpharmaceutical Interventions Implemented by US Cities During 1918–1919 Influenza Pandemic," *Journal of the American Medical Association* 298, no. 6 (2007): 644–54, analyze the effectiveness of interventions in U.S. cities during the Spanish flu pandemic.

117 human society: The advice of the World Health Organization is nonbinding and countries often do not follow the guidelines. S. Gebrekidan, "The World Has a Plan to Fight Coronavirus. Most Countries Are Not Using It," *New York Times*, March 12, 2020.

118 our interconnected world: For examples of network science to control epidemics and design vaccination campaigns, see Zhen Wang et al., "Vaccination and Epidemics in Networked Populations—an Introduction," *Chaos, Solitons, and Fractals* 103 (2017): 177–83; Michael R. Kelly et al., "The Impact of Spatial Arrangements on Epidemic Disease Dynamics and Intervention Strategies," *Journal of Biological Dynamics* 10, no. 1 (2016): 222–49; Matthew J. Keeling, "The Effects of Local Spatial Structure on Epidemiological Invasions," *Proceedings of the Royal Society of London B: Biological Sciences* 266, no. 1421 (1999): 859–67; and Romualdo Pastor-Satorras et al., "Epidemic Processes in Complex Networks," *Reviews of Modern Physics* 87, no. 3 (2015): 925.

5. ONE SIZE FITS NO ONE: MAKE DECISIONS FROM THE BOTTOM UP

121 "an individual": Aristotle, *Politics* (ReadHowYouWant.com, 2006), https://www.readhowyouwant.com/Books/details/11644, p. 58.

121 *The Limits to Growth*: Paul Ehrlich, *The Population Bomb* (New York: Ballantine, 1978); and Donella H. Meadows, Jørgen Randers, and William W. Behrens, *The Limits to Growth: A Report to the Club of Rome* (Washington, D.C.: Potomac Associates, 1972). http://www.donellameadows.org/wp-content/userfiles/Limits-to-Growth-digital-scan-version.pdf

122 "ruin to all":Garrett Hardin, "The Tragedy of the Commons," *Science* 162, no. 3859 (1968): 1244.

122 he reasoned: Hardin, "The Tragedy of the Commons," 1245.

123 use the resource: Hardin, "The Tragedy of the Commons," 1245.

123 "freedom to breed": Hardin, "The Tragedy of the Commons," 1248.

123 "nature's bounty": For example, R. T. Wright and B. J. Nebel, Environmental *Science: Toward a Sustainable Future*, 8th ed. (Upper Saddle River, N.J., 2000), has a section on the tragedy of commons. Quote from R. Lineberry and G. Edwards, *Government in America: People, Politics, and Policy* (Glenview, Ill.: Scott Foresman, 1989), 633.

123 fishing boats: Fisheries and Marine Service, *Policy for Canada's Commercial Fisheries* (Ottawa: Department of the Environment, Government of Canada, 1976), cited in Ralph Matthews, "Federal Licencing Policies for the Atlantic Inshore Fishery and Their Implementation in Newfoundland, 1973–1981," *Acadiensis* 17, no. 2 (1988): 83–108.

123 degradation of the steppe: Dawn Chatty, "The Bedouin in Contemporary Syria: The Persistence of Tribal Authority and Control," *Middle East Journal* 64, no. 1 (2010): 29–49.

123 government ownership: Elinor Ostrom et al., eds., *The Drama of the Commons* (Washington, D.C.: National Academies Press, 2002).

125 "challenging dilemmas": Elinor Ostrom, "A Long Polycentric Journey," *Annual Review of Political Science* 13 (2010): 6.

126 the conclusion: Ostrom, "A Long Polycentric Journey," 10.

126 in what season: See David Feeny et al., "The Tragedy of the Commons: Twenty-Two Years Later," *Human Ecology* 18, no. 1 (1990): 1–19, for criticism of the tragedy of the commons. Hardin later conceded that he had not accounted for difference between open-access and managed commons. Garrett Hardin, "Extensions of 'the Tragedy of the Commons,'" *Science* 280, no. 5364 (1998): 682–83.

127 after one child: M. Levi, "An Interview with Elinor Ostrom," *Annual Reviews Conversations* (2010), https://www.annualreviews.org/userimages/ContentEditor/1326999553977/ElinorOstromTranscript.pdf , p. 8.

127 "birth-control policy": D. Cole and M. McGinnis, eds., *Elinor Ostrom and the Bloomington School of Political Economy*, Vol. 2, *Resource Governance* (Lanham, Md.: Lexington, 2015), xiv.

127 sometimes in comedy: Ostrom et al., *The Drama of the Commons.*

127 guard against tragedy: Robert McC. Netting, "What Alpine Peasants Have in Common: Observations on Communal Tenure in a Swiss Village," in *Case Studies in Human Ecology*, ed. Daniel G. Bates and Susan H. Lees (New York: Springer, 1996), 219–31; and Robert McC. Netting, *Balancing on an Alp: Ecological Change and Continuity in a Swiss Mountain Community* (Cambridge: Cambridge University Press, 1981).

128 threats of violence: Fikret Berkes, "Local-Level Management and the Commons Problem: A Comparative Study of Turkish Coastal Fisheries," *Marine Policy* 10, no. 3 (1986): 215–29.

129 they had little control: Elinor Ostrom, "Do Institutions for Collective Action Evolve?," *Journal of Bioeconomics* 16, no. 1 (2014): 3–30; and Wai Fung Lam, *Governing Irrigation Systems in Nepal: Institutions, Infrastructure, and Collective Action* (Oakland, Calif.: Institute for Contemporary Studies, 1998).

129 imposition from officials: A full list of design principles is found in Elinor Ostrom, *Governing the Commons: The Evolution of Institutions for Collective Action* (Cambridge: Cambridge University Press, 1990), 90.

130 "external authorities": Cole and McGinnis, *Elinor Ostrom and the Bloomington School of Political Economy*, 293.

130 a Canadian official: Quote from John Kearney, "The Transformation of the Bay of Fundy Herring Fisheries, 1976–1978: An Experiment in Fishermen–Government Comanagement," *Atlantic Fisheries and Coastal Communities: Fisheries Decision-making Case Studies* 1984: : 165–203, cited in Robert S. Pomeroy and Fikret Berkes, "Two to Tango: The Role of Government in Fisheries Co-Management," *Marine Policy* 21, no. 5 (1997): 465–80

130 other countries: Pomeroy and Berkes,"Two to Tango"; and William Dillinger, *Decentralization and Its Implications for Urban Service Delivery* (Washington, D.C.: World Bank, 1994).

131 degrees of success: See Jesse C. Ribot, Arun Agrawal, and Anne M. Larson, "Recentralizing While Decentralizing: How National Governments Reappropriate Forest Resources," *World Development* 34, no. 11 (2006): 1864–86, for analysis of case studies in six countries; and Arun Agrawal and Elinor Ostrom, "Collective Action, Property Rights, and Decentralization in Resource Use in India and Nepal," *Politics & Society* 29, no. 4 (2001): 485–514, for case studies of India and Nepal.

131 "policy-makersatalllevels":TheDublinStatementisavailableatwww.wmo.int/pages/prog/hwrp/documents/english/icwedece.html. See Principle No. 2.

131 "can no longer handle itself!": Quote from Elinor Ostrom, "Decentralization and Development: The New Panacea," in *Challenges to Democracy*, ed. Keith Dowding, James Hughes, and Helen Margetts (London: Springer, 2001), 237, 252, and 253.

132 "in the harvest": From Jeremiah 6:6–8, in T. Cheyne et al., eds., *The Holy Bible, Containing the Old and New Testaments: Transalted out of the Original Tongues: And with the Former Translations Diligently Compared and Revised, by His Majesty's Special Command* (London: George Edward Eyre and William Spottiswoode, 1880).

133 searched for food on her own: For self-organized behavior in social animal, see David J. T. Sumpter, *Collective Animal Behavior* (Princeton, N.J.: Princeton University Press, 2010); and Eric Bonabeau, "Social Insect Colonies as Complex Adaptive Systems," *Ecosystems* 1, no. 5 (1998): 437–43. For early works on modeling of self-organized behavior

in ecosystems, see Simon A. Levin, "Ecosystems and the Biosphere as Complex Adaptive Systems," *Ecosystems* 1, no. 5 (1998): 431–36; and Gregoire Nicolis and and Ilya Prigogine, *Self-Organization in Non-Equilibrium Systems* (New York: Wiley, 1977).

134 decide which path to follow: Experiments to determine how ants collectively choose the shortest route are reported in Simon Goss et al., "Self-Organized Shortcuts in the Argentine Ant," *Naturwissenschaften* 76, no. 12 (1989): 579–81; and J.-L. Deneubourg et al., "The Self-Organizing Exploratory Pattern of the Argentine Ant," *Journal of Insect Behavior* 3, no. 2 (1990)159–68. The ant colony optimization algorithm is used to solve many routing problems. Marco Dorigo and Mauro Birattari, "Ant Colony Optimization" in *Encyclopedia of Machine Learning*, ed. C. Sammut and G.I. Webb (New York: Springer, 2011).

134 colonial-era monarchies: Deborah M. Gordon, *Ant Encounters: Interaction Networks and Colony Behavior* (Primers in Complex Systems 1, Princeton, N.J.: Princeton University Press, 2010), describes the changing views of ant colonies.

134 enough offspring: Luke Holman, Stephanie Dreier, and Patrizia d'Ettorre, "Selfish Strategies and Honest Signalling: Reproductive Conflicts in Ant Queen Associations," *Proceedings of the Royal Society of London B: Biological Sciences* 277, no. 1690 (2010): 2007–15.

134 neighbor in front: Emmanuel Hermellin and Fabien Michel, "Complex Flocking Dynamics Without Global Stimulus" (paper presented at the Artificial Life Conference Proceedings 14, Lyon, France, September 4–8, 2017); Michele Ballerini et al., "Interaction Ruling Animal Collective Behavior Depends on Topological Rather Than Metric Distance: Evidence from a Field Study," *Proceedings of the National Academy of Sciences* 105, no. 4 (2008): 1232–37; Federico Cattivelli and Ali H. Sayed, "Self-Organization in Bird Flight Formations Using Diffusion Adaptation" (paper presented at the 2009 3rd IEEE International Workshop on Computational Advances in Multi-Sensor Adaptive Processing [CAMSAP], Aruba, Dutch Antilles, December 13-16, 2009); and Steven J. Portugal et al., "Upwash Exploitation and Downwash Avoidance by Flap Phasing in Ibis Formation Flight," *Nature* 505, no. 7483 (2014): 399.

134 avoid collisions: See Craig W Reynolds, *Flocks, Herds and Schools: A Distributed Behavioral Model*, in *SIGGRAPH '87: Proceedings of the 14th*

Annual Conference on Computer Graphics and Interactive Techniques, ed. Maureen C. Stone (New York: Association for Computing Machinery, 1987), , 25–34, for rules for fish schools.

135 yet to be uncovered: See chap. 18 in Scott Camazine et al., *Self-Organization in Biological Systems* (Princeton, N.J.: Princeton University Press, 2003), on termite mound building. The African termites are in the subfamily *Macrotermitinae.*

136 the activation one: Scott Camazine, "Self-Organizing Systems," *Encyclopedia of Cognitive Science* (Hoboken, N.J.: Wiley,2006); and Camazine et al., *Self-Organization in Biological Systems.*

138 "distribution of the necessaries of life": Adam Smith, *The Wealth of Nations: An Inquiry Into the Nature and Causes of the Wealth of Nations* (Petersfield, U.K.: Harriman House, 2010), 331.

140 "not guaranteed to work well": Elinor Ostrom, "Polycentric Systems for Coping with Collective Action and Global Environmental Change," *Global Environmental Change* 20, no. 4 (2010): 550.

140 into the atmosphere: Svante Arrhenius, "XXXI. On the Influence of Carbonic Acid in the Air Upon the Temperature of the Ground," *London, Edinburgh, and Dublin Philosophical Magazine and Journal of Science* 41, no. 251 (1896): 271.

141 "rapidly propagating mankind": Svante Arrhenius, *Worlds in the Making: The Evolution of the Universe* (New York: Harper & Brothers, 1908), 63.

141 reaching an end: The origin of the phrase "humanity's period of grace" is not clear. It has been used by scientist Johannes Rockstrom, previous director of the Stockholm Resilience Center

141 "address climate change": C. Davenport, "Nations Approve Landmark Climate Accords in Paris," *New York Times*, December 12, 2015.

142 "international agreement": United Nations, "We the Peoples: The Role of the United Nations in the Twenty-First Century. Report of the Secretary-General, a/54/2000, 27 March 2000" (New York: United Nations General Assembly, 2000), 42.

142 "the climate system": United Nations, "United Nations Framework Convention on Climate Change," FCCC/INFORMAL/84 GE.05-62220 (E) 200705 (1992), 4.

144 Hardin would have suggested: Documents and decisions under the United Nations Framework for Climate Change are recorded at https://unfccc.int/documents.

144 "hedge and retreat": The pledge-and-review approach in early days of climate negotiations is discussed in Daniel Bodansky, "On Climate Change: A Commentary," *Yale Journal of International Law* 18, no. 451 (1993): 486; and Steinar Andresen, "International Climate Negotiations: Top-Down, Bottom-up or a Combination of Both?," *International Spectator* 50, no. 1 (2015): 15–30.

145 "level of ambition": United Nations, "Paris Agreement, Chapter XXVII, 7.D Paris Agreement, 12 December 2015" (2015).

145 guided the process: The Paris Agreement is discussed in Robert O. Keohane and Michael Oppenheimer, "Paris: Beyond the Climate Dead End through Pledge and Review?," *Politics and Governance* 4, no. 3 (2016): 142–51; and Andresen, "International Climate Negotiations."

146 "what is behind all of these measures": Quote from Christiana Figueres can be found at https://climateparis.org/Christiana-Figueres-INDCs.

146 banner: We Are Still In! home page, https://www.wearestillin.com.

146 "C40": C40 Cities home page, https://www.c40.org.

147 global concerns can align: Guy J. Abel et al., "Meeting the Sustainable Development Goals Leads to Lower World Population Growth," *Proceedings of the National Academy of Sciences* 113, no. 50 (2016): 14294–99; and Wolfgang Lutz, William P. Butz, and K. C. Samir, eds., *World Population and Human Capital in the Twenty-First Century* (Oxford: Oxford University Press, 2014).

6. CYCLES OF RENEWAL

153 cooperation among countries: Warder Clyde Allee, *Cooperation Among Animals with Human Implications* (New York: Schuman, 1951), 199.

153 "less happy state": Robert M. May, Simon A. Levin, and George Sugihara, "Complex Systems: Ecology for Bankers," *Nature* 451, no. 7181 (2008): 893. This paper followed from a collaboration between the National Research Council and the New York Fed National Research Council, *New Directions for Understanding Systemic Risk: A Report on a Conference Cosponsored by the Federal Reserve Bank of New York and the National Academy of Sciences* (Washington, D.C.: National Academies Press, 2007).

153 national security: Rafe Sagarin, *Learning from the Octopus: How Secrets from Nature Can Help Us Fight Terrorist Attacks, Natural Disasters, and Disease* (New York: Basic Books, 2012).

155 got a foothold: Ramesh Kannan, Charlie M Shackleton, and R. Uma Shaanker, "Reconstructing the History of Introduction and Spread of the Invasive Species, Lantana, at Three Spatial Scales in India," *Biological Invasions* 15, no. 6 (2013): 1287–302.

156 the status quo: Niki Frantzeskaki and Derk Loorbach, "Towards Governing Infrasystem Transitions: Reinforcing Lock-In or Facilitating Change?," *Technological Forecasting and Social Change* 77, no. 8 (2010): 1292–301; Gregory C. Unruh, "Understanding Carbon Lock-In," *Energy Policy* 28, no. 12 (2000): 817–30; and Karen C. Seto et al., "Carbon Lock-In: Types, Causes, and Policy Implications," *Annual Review of Environment and Resources* 41 (2016): 425–52.

156 the system remains locked: Stephen R. Carpenter and William A. Brock, "Adaptive Capacity and Traps," *Ecology and Society* 13, no. 2 (2008): ar40.

157 planets to civilizations: Crawford S. Holling, "Resilience and Stability of Ecological Systems," *Annual Review of Ecology and Systematics* 4, no. 1 (1973): 1–23. Sources for C. S. Holling's work on resilience and panarchy include, among many others, Lance H. Gunderson, *Panarchy: Understanding Transformations in Human and Natural Systems* (Washington, D.C.: Island Press, 2001); Crawford Stanley Holling, "Engineering Resilience Versus Ecological Resilience," in *Engineering Within Ecological Constraints*, ed. P. Schulze (National Academy of Engineering 1996): 31–45 Brian Walker et al., "Resilience, Adaptability and Transformability in Social–Ecological Systems," *Ecology and Society* 9, no. 2 (2004): 5.

158 Tainter writes: Quote from Joseph Tainter, *The Collapse of Complex Societies* (Cambridge: Cambridge University Press, 1988), 119.

158 "marginal return": Quote from Tainter, *The Collapse of Complex Societies*, 120

159 "precise world made of molecules": Quote from Sagarin, *Learning from the Octopus*, 139.

160 "But they didn't": American Society for Cybernetics and H. Von Foerster, *Purposive Systems: Proceedings of the First Annual Symposium of the American Society for Cybernetics* (Washington, D.C.: Spartan, 1969), 2

160 self-regulating feedbacks: Norbert Wiener, *Cybernetics or Control and Communication in the Animal and the Machine*, 2nd ed. (Cambridge, Mass.: MIT Press, 1965).

160 the body and mind: For the history of cybernetics, see Stuart Umpleby, "A Short History of Cybernetics in the United States," *Österreichische Zeitschrift für Geschichtswissenschaften* 19, no. 4 (2008): 28; Kevin Kelly, *Out of Control: The New Biology of Machines, Social Systems, and the Economic World* (London: Hachette UK, 2009); and Robert Vallée, "History of Cybernetics," *Systems Science and Cybernetics*, vol. 3, ed. F. Parra-Luna (Oxford: Eolss, 2009): 22–33.

REFERENCES

Abel, Guy J., Bilal Barakat, K. C. Samir, and Wolfgang Lutz. "Meeting the Sustainable Development Goals Leads to Lower World Population Growth." *Proceedings of the National Academy of Sciences* 113, no. 50 (2016): 14294–99.

Abrams, Peter A. "The Evolution of Predator-Prey Interactions: Theory and Evidence." *Annual Review of Ecology and Systematics* 31, no. 1 (2000): 79–105.

Agrawal, Arun, and Elinor Ostrom. "Collective Action, Property Rights, and Decentralization in Resource Use in India and Nepal." *Politics & Society* 29, no. 4 (2001): 485–514.

Akansu, Ali N. "The Flash Crash: A Review." *Journal of Capital Markets Studies* 1, no. 1 (2017): 89–100.

Aldrich, Eric, Joseph Grundfest, and Gregory Laughlin. "The Flash Crash: A New Deconstruction." SSRN, March 26, 2017. https://ssrn.com/abstract=2721922.

Allee, Warder Clyde. *Cooperation Among Animals with Human Implications*. New York: Schuman, 1951.

Allen, J. P., Mark Nelson, and Abigail Alling. "The Legacy of Biosphere 2 for the Study of Biospherics and Closed Ecological Systems." *Advances in Space Research* 31, no. 7 (2003): 1629–39.

Allen, Toph, Kris A. Murray, Carlos Zambrana-Torrelio, Stephen S. Morse, Carlo Rondinini, Moreno Di Marco, Nathan Breit, Kevin J. Olival, and Peter Daszak. "Global Hotspots and Correlates of Emerging Zoonotic Diseases." *Nature Communications* 8, no. 1 (2017): 1124.

Alou, Maryam Tidjani, Jean-Christophe Lagier, and Didier Raoult. "Diet Influence on the Gut Microbiota and Dysbiosis Related to Nutritional Disorders." *Human Microbiome Journal* 1 (2016): 3–11.

Alter, Harvey J., and Harvey G. Klein. "The Hazards of Blood Transfusion in Historical Perspective." *Blood* 112, no. 7 (2008): 2617–26.

Althoff, David M., Kari A. Segraves, and Marc T. J. Johnson. "Testing for Coevolutionary Diversification: Linking Pattern with Process." *Trends in Ecology & Evolution* 29, no. 2 (2014): 82–89.

American Society for Cybernetics, and H. Von Foerster. *Purposive Systems: Proceedings of the First Annual Symposium of the American Society for Cybernetics*. Washington, D.C.: Spartan, 1969.

Aminov, Rustam I. "A Brief History of the Antibiotic Era: Lessons Learned and Challenges for the Future." *Frontiers in Microbiology* 1 (2010): 134.

Andresen, Steinar. "International Climate Negotiations: Top-Down, Bottom-Up or a Combination of Both?" *International Spectator* 50, no. 1 (2015): 15–30.

Antunes, L. Caetano M., Jun Han, Rosana B.R. Ferreira, Petra Lolić, Christoph H. Borchers, and B. Brett Finlay. "The Effect of Antibiotic Treatment on the Intestinal Metabolome." *Antimicrobial Agents and Chemotherapy* 55, no. 4 (2011): 1494–1503.

Aristotle. *Politics* (ReadHowYouWant.com, 2006). https://www.readhowyouwant.com/Books/details/11644.

Arrhenius, Svante. *Worlds in the Making: The Evolution of the Universe*. New York: Harper & Brothers, 1908.

——. "XXXI. On the Influence of Carbonic Acid in the Air Upon the Temperature of the Ground." *London, Edinburgh, and Dublin Philosophical Magazine and Journal of Science* 41, no. 251 (1896): 237–76.

"Australia's Worst-Ever Wildfires Kill 130." *CBS News*, February 8, 2009.

Avise, John C. "The Real Message from Biosphere 2." *Conservation Biology* 8, no. 2 (1994): 327–29.

Ballerini, Michele, Nicola Cabibbo, Raphael Candelier, Andrea Cavagna, Evaristo Cisbani, Irene Giardina, Vivien Lecomte, et al. "Interaction Ruling Animal Collective Behavior Depends on Topological Rather Than Metric Distance: Evidence from a Field Study." *Proceedings of the National Academy of Sciences* 105, no. 4 (2008): 1232–37.

Barabási, Albert-László, and Réka Albert. "Emergence of Scaling in Random Networks." *Science* 286, no. 5439 (1999): 509–12.

Barabási, Albert-László, and Eric Bonabeau. "Scale-free." *Scientific American* 288, no. 5 (2003): 50–59.

Baran, Paul. "On Distributed Communications: I. Introduction to Distributed Communicaitons Networks." Santa Monica, Calif.: Rand Coporation Memorandum RM-3420-PR, August 1964.

Barquet, Nicolau, and Pere Domingo. "Smallpox: The Triumph Over the Most Terrible of the Ministers of Death." *Annals of Internal Medicine* 127, no. 8, part 1 (1997): 635–42.

Bar-Yam, Yaneer. "Complexity Rising: From Human Beings to Human Civilization, a Complexity Profile." 2002. In *Encyclopedia of Life Support Systems (EOLSS)*, developed under the Auspices of the UNESCO. Oxford: EOLSS Publishers, 2002.

Bascompte, Jordi, Pedro Jordano, Carlos J Melián, and Jens M Olesen. "The Nested Assembly of Plant–Animal Mutualistic Networks." *Proceedings of the National Academy of Sciences* 100, no. 16 (2003): 9383–87.

Basu, Sanjay, Paula Yoffe, Nancy Hills, and Robert H. Lustig. "The Relationship of Sugar to Population-Level Diabetes Prevalence: An Econometric Analysis of Repeated Cross-Sectional Data." *PLoS ONE* 8, no. 2 (2013): e57873.

Bates, J., A. Mansoor, J. Aguilera, A. Gunia, and M. Carlisle. "Firefighters Make Progress Against Fires Raging in California." *Time*, October 25, 2019.

Bäumler, Andreas J., and Vanessa Sperandio. "Interactions Between the Microbiota and Pathogenic Bacteria in the Gut." *Nature* 535, no. 7610 (2016): 85.

Bello, Walden, and Mara Baviera. "Food Wars." *Monthly Review*, July 1, 2009, 17–31.

Berche, P. "Louis Pasteur, from Crystals of Life to Vaccination." *Clinical Microbiology and Infection* 18 (2012): 1–6.

Berkes, Fikret. "Local-Level Management and the Commons Problem: A Comparative Study of Turkish Coastal Fisheries." *Marine Policy* 10, no. 3 (1986): 215–29.

Berkes, Fikret, and Dyanna Jolly. "Adapting to Climate Change: Social-Ecological Resilience in a Canadian Western Arctic Community." *Conservation Ecology* 5, no. 2 (2002): art18.

Bernhofen, Daniel M., Zouheir El-Sahli, and Richard Kneller. "Estimating the Effects of the Container Revolution on World Trade." *Journal of International Economics* 98 (2016): 36–50.

Beynon-Davies, Paul. "Informatics and the Inca." *International Journal of Information Management* 27, no. 5 (2007): 306–18.

"Bill Clinton: 'We Blew It' on Global Food." *CBS News*, October 23, 2008.

Blaser, Martin J. "Antibiotic Use and Its Consequences for the Normal Microbiome." *Science* 352, no. 6285 (2016): 544–45.

Bodansky, Daniel. "On Climate Change: A Commentary." *Yale Journal of International Law* 18, no. 451 (1993): 453–558.

Bogoch, Isaac I., Oliver J. Brady, Moritz U. G. Kraemer, Matthew German, Marisa I. Creatore, Manisha A. Kulkarni, John S. Brownstein, *et al.* "Anticipating the International Spread of Zika Virus from Brazil." *Lancet* 387, no. 10016 (2016): 335–36.

Bohstedt, J. "Food Riots and the Politics of Provisions in World History." IDS Working Paper No. 444. Brighton, UK: Institute of Development Studies, 2014.

Bohstedt, John. *The Politics of Provisions: Food Riots, Moral Economy, and Market Transition in England, c. 1550–1850*. Farnham, U.K.: Ashgate, 2013.

Bonabeau, Eric. "Social Insect Colonies as Complex Adaptive Systems." *Ecosystems* 1, no. 5 (1998): 437–43.

Boroditsky, Lera. "How Language Shapes Thought." *Scientific American* 304, no. 2 (2011): 62–65.

Bouët, Antoine, and David Laborde Debucquet. "Food Crisis and Export Taxation: Revisiting the Adverse Effects of Noncooperative Aspect of Trade Policies." In *Food Price Volatility and Its Implications for Food Security and Policy*, edited by Matthias Kalkuhl, Joachim von Braun, and Maximo Torero, 167–79. New York: Springer, 2016.

Bowman, David, and Ross A. Bradstock. "Australia Needs a National Fire Inquiry—These Are the 3 Key Areas It Should Deliver In." *The Conversation*, January 22, 2020.

Brand, S. "Founding Father: *Wired* Legends." *Wired*, March 1, 2001.

Cahir, Fred, Sarah McMaster, Ian Clark, Rani Kerin, and Wendy Wright. "Winda Lingo Parugoneit or Why Set the Bush [on] Fire? Fire and Victorian Aboriginal People on the Colonial Frontier." *Australian Historical Studies* 47, no. 2 (2016): 225–40.

Camazine, Scott. "Self-Organizing Systems." *Encyclopedia of Cognitive Science*. Hoboken, N.J.: Wiley, 2006.

Camazine, Scott, Jean-Louis Deneubourg, Nigel R. Franks, James Sneyd, Eric Bonabeau, and Guy Theraula. *Self-Organization in Biological Systems*. Princeton, N.J.: Princeton University Press, 2003.

Carpenter, Stephen R., and William A. Brock. "Adaptive Capacity and Traps." *Ecology and Society* 13, no. 2 (2008): ar40.

Carrington, D. "Arctic Stronghold of World's Seeds Flooded After Permafrost Melts." *The Guardian*, May 19, 2017.

Carroll, Lewis. *Through the Looking Glass and What Alice Found There*. Chicago: Rand McNally, 1917.

Carroll, Sean B. "Chance and Necessity: The Evolution of Morphological Complexity and Diversity." *Nature* 409, no. 6823 (2001): 1102.

Cattivelli, Federico, and Ali H. Sayed. "Self-Organization in Bird Flight Formations Using Diffusion Adaptation." Paper presented at the 2009 3rd IEEE International Workshop on Computational Advances in Multi-Sensor Adaptive Processing (CAMSAP), Aruba, Dutch Antilles, December 13–16, 2009.

Centeno, Miguel A., Manish Nag, Thayer S. Patterson, Andrew Shaver, and A. Jason Windawi. "The Emergence of Global Systemic Risk." *Annual Review of Sociology* 41 (2015): 65–85.

Chancey, Caren, Andriyan Grinev, Evgeniya Volkova, and Maria Rios. "The Global Ecology and Epidemiology of West Nile Virus." *BioMed Research International* 2015: ar376230. https://doi.org/10.1155/2015/376230.

Charlton, L. "Smokey Bear Dies in Retirement." *New York Times*, November 10, 1976.

Chase-Dunn, Christopher, Yukio Kawano, and Benjamin D. Brewer. "Trade Globalization Since 1795: Waves of Integration in the World-System." *American Sociological Review* 65, no. 1 (2000): 77–95.

Chatty, Dawn. "The Bedouin in Contemporary Syria: The Persistence of Tribal Authority and Control." *Middle East Journal* 64, no. 1 (2010): 29–49.

Cheyne, T., R. Clarke, S. Driver, A. Goodwin, and W. Sanday, eds. *The Holy Bible, Containing the Old and New Testaments: Transalted out of the Original Tongues: And with the Former Translations Diligently Compared and Revised, by His Majesty's Special Command*. London: George Edward Eyre and William Spottiswoode, 1880.

Chin, H. F. "Don't Let It Go! Saving Endangered Plants Through Seed Storage." *UTAR Agriculture Science Journal* 2, no. 3 (2016): 57–60.

Cohen, Joel E., and David Tilman. "Biosphere 2 and Biodiversity—The Lessons So Far." *Science* 274, no. 5290 (1996): 1150–51.

Cole, D., and M. McGinnis, eds. *Elinor Ostrom and the Bloomington School of Political Economy.* Vol. 2, *Resource Governance*. Lanham, Md.: Lexington, 2015.

Cole, S. "The Strange History of Steve Bannon and the Biosphere2 Experiment." *Motherboard*, November 15, 2016.

Congressional Budget Office. "The Gatt Negotiations and U.S. Trade Policy." Washington, D.C.: Congress of the United States, 1987.

Corinto, G. L. "Nikolai Vavilov's Centers of Origin of Cultivated Plants with a View to Conserving Agricultural Biodiversity." *Human Evolution* 29, no. 4 (2014): 285–301.

Corson, Francis. "Fluctuations and Redundancy in Optimal Transport Networks." *Physical Review Letters* 104, no. 4 (2010): 048703.

Costello, Mark J, Robert M May, and Nigel E Stork. "Can We Name Earth's Species Before They Go Extinct?." *Science* 339, no. 6118 (2013): 413–16.

Cowling, B., and W. W. Lim. "They've Containted the Coronavirus. Here's How." *New York Times*, March 13, 2020.

Cox, J., and M. Bloom. "The Market Triggered a 'Circuit Breaker' That Kept Stocks from Falling Through the Floor. Here's What You Need to Know." *CNBC*, March 9, 2020.

Crane, P. *Gingko: The Tree That Time Forgot*. New Haven, Conn.: Yale University Press, 2013.

Cremer, S. "Social Immunity in Insects." *Current Biology* 29, no. 11 (2019): R425–R473.

Cremer, Sylvia, and Michael Sixt. "Analogies in the Evolution of Individual and Social Immunity." *Philosophical Transactions of the Royal Society of London B: Biological Sciences* 364, no. 1513 (2008): 129–42.

Darwin, Charles. *The Descent of Man and Selection in Relation to Sex*. Vol. 1. London: Murray, 1888.

Davenport, C. "Nations Approve Landmark Climate Accords in Paris." *New York Times*, December 12, 2015.

Dawkins, Richard, and John Richard Krebs. "Arms Races Between and Within Species." *Proceedings of the Royal Society of London B: Biological Sciences* 205, no. 1161 (1979): 489–511.

DeFries, Ruth. *The Big Ratchet: How Humanity Thrives in the Face of Natural Crisis*. New York: Basic Books, 2014.

DeFries, R., O. Edenhofer, A. Halliday, G. Heal, T. Lenton, M. Puma, J. Rising, et al. "The Missing Economic Risks in Assessments of Climate

Change Impacts." Policy Insight. London: Grantham Research Institute on Climate Change and the Environment, September 2019.

Demandt, Alexander. *Der Fall Roms*. Munich: Beck, 1984.

Deneubourg, J.-L., Serge Aron, Simon Goss, and Jacques M. Pasteels. "The Self-Organizing Exploratory Pattern of the Argentine Ant." *Journal of Insect Behavior* 3, no. 2 (1990): 159–68.

Deria, A., Z. Jezek, K. Markvart, P. Carrasco, and J. Weisfeld. "The World's Last Endemic Case of Smallpox: Surveillance and Containment Measures." *Bulletin of the World Health Organization* 58, no. 2 (1980): 279.

Dillinger, William. *Decentralization and Its Implications for Urban Service Delivery*. Washington, D.C.: World Bank, 1994.

DiMaio, Daniel. "Thank You, Edward. Merci, Louis." *PLoS Pathogens* 12, no. 1 (2016): e1005320.

Dods, Roberta Robin. "The Death of Smokey Bear: The Ecodisaster Myth and Forest Management Practices in Prehistoric North America." *World Archaeology* 33, no. 3 (2002): 475–87.

Doerr, Stefan H., and Cristina Santín. "Global Trends in Wildfire and Its Impacts: Perceptions Versus Realities in a Changing World." *Philosophical Transactions of the Royal Society of London B: Biological Sciences* 371, no. 1696 (2016): 20150345.

Donihue, Colin M., Anthony Herrel, Anne-Claire Fabre, Ambika Kamath, Anthony J. Geneva, Thomas W. Schoener, Jason J. Kolbe, and Jonathan B. Losos. "Hurricane-Induced Selection on the Morphology of an Island Lizard." *Nature* 560 (2018): 88–91.

Donovan, Geoffrey H., and Thomas C. Brown. "Be Careful What You Wish For: The Legacy of Smokey Bear." *Frontiers in Ecology and the Environment* 5, no. 2 (2007): 73–79.

Dorigo, Marco, and Mauro Birattari. "Ant Colony Optimization," in *Encyclopedia of Machine Learning*, ed. C. Sammut and G. I. Webb (New York: Springer, 2011).

Doucleff, Michaeleen. "Last Person To Get Smallpox Dedicated His Life To Ending Polio." *NPR Shots*, July 31, 2013. www.npr.org/sections/health-shots/2013/07/31/206947581/last-person-to-get-smallpox-dedicated-his-life-to-ending-polio.

Downer, John. *When Failure Is an Option: Redundancy, Reliability and Regulation in Complex Technical Systems*. London: Centre for Analysis of Risk and Regulation, London School of Economics and Political Science, 2009.

Doyle, L. "Starving Haitians Riot as Food Prices Soar." *The Independent*, April 9, 2008.

Dunning, N., T. Beach, and S. Luzzadder-Beach. "Kax and Kol: Collapse and Resilience in Lowland Maya Civilization." *Proceedings of the National Academy of Sciences* 109, no. 10 (2012): 3652–57.

Dutfield, Graham. "Opinion: Why Traditional Knowledge Is Important in Drug Discovery." *Future Medicinal Chemistry* 2, no. 9 (2010): 1405–409.

Ebert, Dieter. "Host–Parasite Coevolution: Insights from the Daphnia–Parasite Model System." *Current Opinion in Microbiology* 11, no. 3 (2008): 290–301.

Ehrlich, P. *The Population Bomb*. New York: Ballantine, 1978.

Elahi, Shirin. "Here Be Dragons . . . Exploring the 'Unknown Unknowns.'" *Futures* 43, no. 2 (2011): 196–201.

Ercsey-Ravasz, Mária, Zoltán Toroczkai, Zoltán Lakner, and József Baranyi. "Complexity of the International Agro-Food Trade Network and Its Impact on Food Safety." *PLoS ONE* 7, no. 5 (2012): e37810.

Esquinas-Alcázar, José. "Protecting Crop Genetic Diversity for Food Security: Political, Ethical and Technical Challenges." *Nature Reviews Genetics* 6, no. 12 (2005): 946.

Evans, Nicholas. *Dying Words: Endangered Languages and What They Have to Tell Us*. Chichester, U.K.: Wiley-Blackwell, 2010.

Fabricant, Daniel S., and Norman R. Farnsworth. "The Value of Plants Used in Traditional Medicine for Drug Discovery." *Environmental Health Perspectives* 109, no. S1 (2001): 69.

Feeny, David, Fikret Berkes, Bonnie J. McCay, and James M. Acheson. "The Tragedy of the Commons: Twenty-Two Years Later." *Human Ecology* 18, no. 1 (1990): 1–19.

Fenner, Frank. "Global Eradication of Smallpox." *Reviews of Infectious Diseases* 4, no. 5 (1982): 916–30.

Fenner, Frank, D. A. Henderson, I. Arita, Z. Jezek, and I. Ladnyi. "*Smallpox and Its Eradication*." Geneva: World Health Organization, 1988.

Ferguson, Niall. "Complexity and Collapse: Empires on the Edge of Chaos." *Foreign Affairs*, March/April 2010, 18–32.

Fischer, A. Paige, Thomas A. Spies, Toddi A. Steelman, Cassandra Moseley, Bart R. Johnson, John D. Bailey, Alan A. Ager, et al. "Wildfire Risk as a Socioecological Pathology." *Frontiers in Ecology and the Environment* 14, no. 5 (2016): 276–84.

Fisheries and Marine Service. *Policy for Canada's Commercial Fisheries.* Ottawa: Department of the Environment, Government of Canada, 1976.

Food and Agriculture Organization. *The State of the World's Plant Genetic Resources for Food and Agriculture.* Rome: Food and Agriculture Organization, 1997.

Fowler, C. *"Seeds on Ice.* Westport, Conn.:" Prospecta Press, 2016.

Fox, Douglas. "What Sparked the Cambrian Explosion"?" *Nature News* 530, no. 7590 (2016): 268.

Frantzeskaki, Niki, and Derk Loorbach. "Towards Governing Infrasystem Transitions: Reinforcing Lock-In or Facilitating Change?" *Technological Forecasting and Social Change* 77, no. 8 (2010): 1292–301.

"Frederick Erskine Olmsted." *Journal of Forestry* 23, no. 4 (1925): 338–39.

Fuller, T. "Thailand Flooding Crippled Hard-Drive Suppliers." *New York Times,* November 6, 2011.

Galinsky, Karl. *Classical and Modern Interactions: Postmodern Architecture, Multiculturalism, Decline, and Other Issues.* Austin, Tex.: University of Texas Press, 1992.

Galluzzi, Gea, Michael Halewood, Isabel López Noriega, and Ronnie Vernooy. "Twenty-Five Years of International Exchanges of Plant Genetic Resources Facilitated by the CGIAR Genebanks: A Case Study on Global Interdependence." *Biodiversity and Conservation* 25, no. 8 (2016): 1421–46.

Gaupp, Franziska, Georg Pflug, Stefan Hochrainer-Stigler, Jim Hall, and Simon Dadson. "Dependency of Crop Production Between Global Breadbaskets: A Copula Approach for the Assessment of Global and Regional Risk Pools." *Risk Analysis* 37, no. 11 (2017): 2212–28.

Gebrekidan, S. "The World Has a Plan to Fight Coronavirus. Most Countries Are Not Using It." *New York Times,* March 12, 2020.

Gephart, Jessica A., Elena Rovenskaya, Ulf Dieckmann, Michael L. Pace, and Åke Brännström. "Vulnerability to Shocks in the Global Seafood Trade Network." *Environmental Research Letters* 11, no. 3 (2016): 035008.

Gepts, Paul. "Plant Genetic Resources Conservation and Utilization." *Crop Science* 46, no. 5 (2006): 2278–92.

Gettleman, J. "Monkey in Kenya Survives After Setting Off Nationwide Blackout." *New York Times,* June 8, 2016.

Gibbon, E. *The History of the Decline and Fall of the Roman Empire.* [London?]: W. Strahan and T. Cadell, 1776.

Goldsmith, E., N. Hildyard, P. Bunyard, and P. McCully. "Cakes and Caviar? The Dunkel Draft and Third World Agriculture." *Ecologist* 23, no. 6 (1993): 219–22.

Gordon, Deborah M. *Ant Encounters: Interaction Networks and Colony Behavior*. Primers in Complex Systems 1. Princeton, N.J.: Princeton University Press, 2010.

Goss, Simon, Serge Aron, Jean-Louis Deneubourg, and Jacques Marie Pasteels. "Self-Organized Shortcuts in the Argentine Ant." *Naturwissenschaften* 76, no. 12 (1989): 579–81.

Guarner, Francisco. "Decade in Review—Gut Microbiota: The Gut Microbiota Era Marches On." *Nature Reviews Gastroenterology and Hepatology* 11, no. 11 (2014): 647–49.

Gunderson, Lance H. *Panarchy: Understanding Transformations in Human and Natural Systems*. Washington, D.C.: Island Press, 2001.

"Gunnison and the Great Influenza." *Gunnison Country Times* (Gunnison, Colo.), January 18, 2018.

Haldon, John, Lee Mordechai, Timothy P. Newfield, Arlen F. Chase, Adam Izdebski, Piotr Guzowski, Inga Labuhn, and Neil Roberts. "History Meets Palaeoscience: Consilience and Collaboration in Studying Past Societal Responses to Environmental Change." *Proceedings of the National Academy of Sciences* 115, no. 13 (2018): 3210–3218.

Hamilton, Margaret H., and William R. Hackler. "Universal Systems Language: Lessons Learned from Apollo." *Computer* 41, no. 12 (2008): 34–43.

Hardin, Garrett. "Extensions of" 'The Tragedy of the Commons".' " *Science* 280, no. 5364 (1998): 682–83.

——. "The Tragedy of the Commons." *Science* 162, no. 3859 (1968): 1243–48.

Harlan, Jack R. "Genetics of Disaster 1." *Journal of Environmental Quality* 1, no. 3 (1972): 212–15.

Hassan, Zuhair A. "Agreement on Agriculture in the Uruguay Round of GATT: From Punta Del Este to Marrakesh." *Agricultural Economics* 15, no. 1 (1996): 29–46.

Hawthorne, Matthew J., and Dewayne E. Perry. "Applying Design Diversity to Aspects of System Architectures and Deployment Configurations to Enhance System Dependability." Paper presented at the DSN 2004 Workshop on Architecting Dependable Systems, Florence, Italy, June 28–July 1, 2004 (DSN-WADS 2004).

Headey, Derek, and Shenggen Fan. "Anatomy of a Crisis: The Causes and Consequences of Surging Food Prices." *Agricultural Economics* 39, no. S1 (2008): 375–91.

Heath, T., T. Telford, and H. Long. "Dow Plunges 10 Percent Despite Fed Lifeline as Coronavirus Panic Grips Investors." *Washington Post*, March 12, 2020.

Heckman, Daniel S., David M. Geiser, Brooke R. Eidell, Rebecca L. Stauffer, Natalie L. Kardos, and S. Blair Hedges. "Molecular Evidence for the Early Colonization of Land by Fungi and Plants." *Science* 293, no. 5532 (2001): 1129–33.

Hedges, S. Blair, Jaime E. Blair, Maria L. Venturi, and Jason L. Shoe. "A Molecular Timescale of Eukaryote Evolution and the Rise of Complex Multicellular Life." *BMC Evolutionary Biology* 4, no. 1 (2004): 2.

Heiman, Mark L., and Frank L. Greenway. "A Healthy Gastrointestinal Microbiome Is Dependent on Dietary Diversity." *Molecular Metabolism* 5, no. 5 (2016): 317–20.

Henderson, D. A., and Isao Arita. "The Smallpox Threat: A Time to Reconsider Global Policy." *Biosecurity and Bioterrorism: Biodefense Strategy, Practice, and Science* 12, no. 3 (2014): 117–21.

Henderson, D. A., and Petra Klepac. "Lessons from the Eradication of Smallpox: An Interview with DA Henderson." *Philosophical Transactions of the Royal Society of London B: Biological Sciences* 368, no. 1623 (2013): ar20130113.

Henderson, Donald A. "The Eradication of Smallpox." *Scientific American* 235, no. 4 (1976): 25–33.

——. "Lessons from the Eradication Campaigns." *Vaccine* 17 (1999): S53–S55.

——. "Principles and Lessons from the Smallpox Eradication Programme." *Bulletin of the World Health Organization* 65, no. 4 (1987): 535.

Hermellin, Emmanuel, and Fabien Michel. "Complex Flocking Dynamics Without Global Stimulus." Paper presented at the Artificial Life Conference Proceedings 14, New York, July 30–August 2, 2014.

Hinman, Alan R. "Measles and Rubella Eradication." *Vaccine* 36, no. 1 (2018): 1–3.

Holling, Crawford S. "Resilience and Stability of Ecological Systems." *Annual Review of Ecology and Systematics* 4, no. 1 (1973): 1–23.

Holling, Crawford Stanley. "Engineering Resilience Versus Ecological Resilience." In *Engineering Within Ecological Constraints*, ed. P. Schulz (Washington, D.C.: National Academies Press, 1996): 31–45.

Holman, Luke, Stephanie Dreier, and Patrizia d'Ettorre. "Selfish Strategies and Honest Signalling: Reproductive Conflicts in Ant Queen Associations." *Proceedings of the Royal Society of London B: Biological Sciences* 277, no. 1690 (2010): 2007–15.

Hotez, Peter J., Jeff Bethony, Maria Elena Bottazzi, Simon Brooker, and Paulo Buss. "Hookworm": 'The Great Infection of Mankind".' " *PLoS Medicine* 2, no. 3 (2005): e67.

Hung, Lee Shiu. "The SARS Epidemic in Hong Kong: What Lessons Have We Learned"?" *Journal of the Royal Society of Medicine* 96, no. 8 (2003): 374–78.

Joffard, Nina, François Massol, Matthias Grenié, Claudine Montgelard, and Bertrand Schatz. "Effect of Pollination Strategy, Phylogeny and Distribution on Pollination Niches of Euro-Mediterranean Orchids." *Journal of Ecology* 107, no. 1 (2019): 478–90.

Johnson, Elizabeth L., Stacey L. Heaver, William A. Walters, and Ruth E. Ley. "Microbiome and Metabolic Disease: Revisiting the Bacterial Phylum Bacteroidetes." *Journal of Molecular Medicine* 95, no. 1 (2017): 1–8.

Johnston, Basil H. "One Generation from Extinction." *Native Writers and Canadian Writing* 1990: 10–15.

Jolliffe, D. M. "A History of the Use of Arsenicals in Man." *Journal of the Royal Society of Medicine* 86, no. 5 (1993): 287.

Jones, James Holland, and Mark S. Handcock. "An Assessment of Preferential Attachment as a Mechanism for Human Sexual Network Formation." *Proceedings of the Royal Society of London B: Biological Sciences* 270, no. 1520 (2003): 1123–28.

Jones, Kate E., Nikkita G. Patel, Marc A. Levy, Adam Storeygard, Deborah Balk, John L. Gittleman, and Peter Daszak. "Global Trends in Emerging Infectious Diseases." *Nature* 451, no. 7181 (2008): 990.

Jones, Rhys. "Fire-Stick Farming." *Fire Ecology* 8, no. 3 (2012): 3–8.

Kannan, Ramesh, Charlie M. Shackleton, and R. Uma Shaanker. "Reconstructing the History of Introduction and Spread of the Invasive Species, Lantana, at Three Spatial Scales in India." *Biological Invasions* 15, no. 6 (2013): 1287–302.

Karanja, S. "Monkey Tripped Transformer at Power Station Causing Massive Blackout, Kengen Says." *Daily Nation*, June 8, 2016.

Katifori, Eleni, and Marcelo O. Magnasco. "Quantifying Loopy Network Architectures." *PLoS ONE* 7, no. 6 (2012): e37994.

Katifori, Eleni, Gergely J. Szöllősi, and Marcelo O. Magnasco. "Damage and Fluctuations Induce Loops in Optimal Transport Networks." *Physical Review Letters* 104, no. 4 (2010): 048704.

Kauffman, Stuart. *At Home in the Universe: The Search for the Laws of Self-Organization and Complexity*. New York: Oxford University Press, 1996.

Kearney, John. "The Transformation of the Bay of Fundy Herring Fisheries, 1976–1978: An Experiment in Fishermen–Government Comanagement." *Atlantic Fisheries and Coastal Communities: Fisheries Decision-making Case Studies* 1984: 165–203.

Keeling, Matthew J. "The Effects of Local Spatial Structure on Epidemiological Invasions." *Proceedings of the Royal Society of London B: Biological Sciences* 266, no. 1421 (1999): 859–67.

Keleman, Alder, and Hugo García Rañó. "The Mexican Tortilla Crisis of 2007: The Impacts of Grain-Price Increases on Food-Production Chains." *Development in Practice* 21, no. 4–5 (2011): 550–65.

Kelly, Kevin. *Out of Control: The New Biology of Machines, Social Systems, and the Economic World*. London: Hachette UK, 2009.

Kelly, Michael R., Jr., Joseph H. Tien, Marisa C. Eisenberg, and Suzanne Lenhart. "The Impact of Spatial Arrangements on Epidemic Disease Dynamics and Intervention Strategies." *Journal of Biological Dynamics* 10, no. 1 (2016): 222–49.

Keohane, Robert O., and Michael Oppenheimer. "Paris: Beyond the Climate Dead End Through Pledge and Review?" *Politics and Governance* 4, no. 3 (2016): 142–51.

Kharrazi, Ali, Elena Rovenskaya, and Brian D. Fath. "Network Structure Impacts Global Commodity Trade Growth and Resilience." *PLoS ONE* 12, no. 2 (2017): e0171184.

Khoury, Colin K., Harold A. Achicanoy, Anne D. Bjorkman, Carlos Navarro-Racines, Luigi Guarino, Ximena Flores-Palacios, Johannes M. M. Engels, et al. "Origins of Food Crops Connect Countries Worldwide." *Proceedings of the Royal Society of London B: Biological Sciences* 283, no. 1832 (2016): 20160792.

Khoury, Colin K., Anne D. Bjorkman, Hannes Dempewolf, Julian Ramirez-Villegas, Luigi Guarino, Andy Jarvis, Loren H. Rieseberg, and Paul C. Struik. "Increasing Homogeneity in Global Food Supplies and the Implications for Food Security." *Proceedings of the National Academy of Sciences* 111, no. 11 (2014): 4001–4006.

Kilian, Lutz. "Explaining Fluctuations in Gasoline Prices: A Joint Model of the Global Crude Oil Market and the US Retail Gasoline Market." *The Energy Journal* 31, no. 2 (2010).

Kilpatrick, A. Marm, Andrew D. M. Dobson, Taal Levi, Daniel J. Salkeld, Andrea Swei, Howard S. Ginsberg, Anne Kjemtrup, et al. "Lyme Disease Ecology in a Changing World: Consensus, Uncertainty and Critical Gaps for Improving Control." *Philosophical Transactions of the Royal Society of London B: Biological Sciences* 372, no. 1722 (2017): 20160117.

Kimmerer, Robin. *Braiding Sweetgrass: Indigenous Wisdom, Scientific Knowledge and the Teachings of Plants*. Minneapolis, Minn.: Milkweed Editions, 2013.

Kirilenko, Andrei, Albert S. Kyle, Mehrdad Samadi, and Tugkan Tuzun. "The Flash Crash: High-Frequency Trading in an Electronic Market." *Journal of Finance* 72, no. 3 (2017): 967–98.

Kleinfeld, Judith S. "Six Degrees of Separation: Urban Myth?" *Psychology Today* 35, no. 2 (2002): 74.

Klimek, Peter, Michael Obersteiner, and Stefan Thurner. "Systemic Trade Risk of Critical Resources." *Science Advances* 1, no. 10 (2015): e1500522.

Kwai, I. "Devastating Australia Bush Fire Destroys Scores of Homes." *New York Times*, March 19, 2018.

Lam, Wai Fung. *Governing Irrigation Systems in Nepal: Institutions, Infrastructure, and Collective Action*. Oakland, Calif.: Institute for Contemporary Studies, 1998.

Lama, G. de."Biosphere 2 Proves a Hothouse for Trouble: Project Yields a Crop of Rivalry, Confusion." *Chicago Tribune*, April 16, 1994.

Landau, Martin. "Redundancy, Rationality, and the Problem of Duplication and Overlap." *Public Administration Review* 29, no. 4 (1969): 346–58.

Larsen, Brendan B., Elizabeth C. Miller, Matthew K. Rhodes, and John J. Wiens. "Inordinate Fondness Multiplied and Redistributed: The Number of Species on Earth and the New Pie of Life." *Quarterly Review of Biology* 92, no. 3 (2017): 229–65.

"Last Speaker of Majhi Language Dead." *Times of India*, July 22, 2016.

Latora, Vito, and Massimo Marchiori. "Efficient Behavior of Small-World Networks." *Physical Review Letters* 87, no. 19 (2001): 198701.

Leiner, Barry M., Vinton G. Cerf, David D. Clark, Robert E. Kahn, Leonard Kleinrock, Daniel C. Lynch, Jon Postel, Larry G. Roberts, and Stephen Wolff. "A Brief History of the Internet." *ACM SIGCOMM Computer Communication Review* 39, no. 5 (2009): 22–31.

Lenton, Tim, Andrew Watson, and Andrew J. Watson. *Revolutions That Made the Earth*. Oxford: Oxford University Press, 2011.

Leopold, Aldo. *A Sand County Almanac, and Sketches Here and There*. Outdoor Essays & Reflections. New York: Oxford University Press, 1989. Originally published in 1949.

Leopold, Aldo Starker. *Wildlife Management in the National Parks* (U.S. National Park Service, 1963). https://www.nps.gov/parkhistory/online_books/admin_policies/policy4-leopold.htm.

Levi, M. "An Interview with Elinor Ostrom"." *Annual Reviews Conversations* (2010). https://www.annualreviews.org/userimages/ContentEditor/1326999553977/ElinorOstromTranscript.pdf.

Levin, Simon A. "Ecosystems and the Biosphere as Complex Adaptive Systems." *Ecosystems* 1, no. 5 (1998): 431–36.

Ley, Ruth E., Peter J. Turnbaugh, Samuel Klein, and Jeffrey I. Gordon. "Microbial Ecology: Human Gut Microbes Associated with Obesity." *Nature* 444, no. 7122 (2006): 1022.

Liljeros, Fredrik, Christofer R. Edling, Luis A. Nunes Amaral, H. Eugene Stanley, and Yvonne Åberg. "The Web of Human Sexual Contacts." *Nature* 411, no. 6840 (2001): 907–908.

Lineberry, R., and G. Edwards. *Government in America: People, Politics, and Policy*. Glenview, Ill.: Scott Foresman, 1989.

Littlewood, Bev, Peter Popov, and Lorenzo Strigini. "Design Diversity: An Update from Research on Reliability Modelling." In *Aspects of Safety Management*, edited by Felix Redmill and Tom Anderson, 139–54. London: Springer, 2001.

Liu, L., X.-Y. Zhao, Q.-B. Tang, C.-L Lei, and Q.-Y. Huang. "The Mechanisms of Social Immunity Against Fungal Infections of Eusocial Insects." *Toxins* 11, no. 244 (2019): 1–21.

Locey, Kenneth J, and Jay T Lennon. "Scaling Laws Predict Global Microbial Diversity." *Proceedings of the National Academy of Sciences* 113, no. 21 (2016): 5970–75.

Loh, Jonathan, and David Harmon. "Biocultural Diversity: Threatened Species, Endangered Languages." Zeist, The Netherlands: WWF Netherlands, June 8, 2014.

Lutz, Wolfgang, William P. Butz, and K. C. Samir, eds. *World Population and Human Capital in the Twenty-First Century*. Oxford: Oxford University Press, 2014.

Malhotra, K. "The Uruguay Round of Gatt, the World Trade Organization and Small Farmers." Paper prepared for the Regional Conference on MonoCultural Cropping in Southeast Asia: Social/Environmental Impacts and Sustainable Alternatives, Songkhla, Thailand, June 3–6, 1996.

Malkin, E. "Thousands in Mexico City Protest Rising Food Prices." *New York Times*, February 1, 2007.

Marchand, Philippe, Joel A. Carr, Jampel Dell'Angelo, Marianela Fader, Jessica A. Gephart, Matti Kummu, Nicholas R. Magliocca, et al. "Reserves and Trade Jointly Determine Exposure to Food Supply Shocks." *Environmental Research Letters* 11, no. 9 (2016): 095009.

Marineli, Filio, Gregory Tsoucalas, Marianna Karamanou, and George Androutsos. "Mary Mallon (1869–1938) and the History of Typhoid Fever." *Annals of Gastroenterology* 26, no. 2 (2013): 132.

Markel, H., H. Lipman, J. Navarro, A. Sloan, J. Michalsen, A. Stern, and M. Cetron. "Nonpharmaceutical Interventions Implemented by US Cities During 1918–1919 Influenza Pandemic." *Journal of the American Medical Association* 298, no. 6 (2007): 644–54.

Markowitz, Harry. "Portfolio Selection." *Journal of Finance* 7, no. 1 (1952): 77–91.

Martin, Will, and Kym Anderson. *Export Restrictions and Price Insulation During Commodity Price Booms*. Washington, D.C.: World Bank, 2011.

Matthews, Ralph. "Federal Licencing Policies for the Atlantic Inshore Fishery and Their Implementation in Newfoundland, 1973–1981." *Acadiensis* 17, no. 2 (1988): 83–108.

May, Robert M., Simon A. Levin, and George Sugihara. "Complex Systems: Ecology for Bankers." *Nature* 451, no. 7181 (2008): 893.

McCartney, Robert J. "Mexico to Lower Trade Barriers, Join GATT." *Washington Post*, November 26, 1985.

McConnell, Joseph R., Andrew I. Wilson, Andreas Stohl, Monica M. Arienzo, Nathan J. Chellman, Sabine Eckhardt, Elisabeth M. Thompson, et al. "Lead Pollution Recorded in Greenland Ice Indicates European Emissions Tracked Plagues, Wars, and Imperial Expansion During Antiquity." *Proceedings of the National Academy of Sciences* 115, no. 22 (2018): 5726–5731.

McMillan, Robert. "Her Code Got Humans on the Moon—and Invented Software Itself." *Wired*. https://www.wired.com/2015/10/margaret-hamilton-nasa-apollo.

McWhorter, John H. *The Language Hoax: Why the World Looks the Same in Any Language*. New York: Oxford University Press, 2014.

Meadows, Donella H., Jørgen Randers, and William W. Behrens III. *The Limits to Growth: A Report to the Club of Rome* (Washington, D.C.: Potomac Associates, 1972). http://www.donellameadows.org/wp-content/userfiles/Limits-to-Growth-digital-scan-version.pdf.

Medina, J., L. Stack, and J. Bromwich. "Tens of Thousands Evacuate as Southern California Fires Spread." *New York Times*, December 5, 2017.

Menkveld, Albert J., and Bart Yueshen. "The Flash Crash: A Cautionary Tale About Highly Fragmented Markets." *Management Science* 65 (2019): 4470–4488.

Middleton, Guy D. "Nothing Lasts Forever: Environmental Discourses on the Collapse of Past Societies." *Journal of Archaeological Research* 20, no. 3 (2012): 257–307.

Mills, Daniel B., Lewis M. Ward, CarriAyne Jones, Brittany Sweeten, Michael Forth, Alexander H. Treusch, and Donald E. Canfield. "Oxygen Requirements of the Earliest Animals." *Proceedings of the National Academy of Sciences* 111, no. 11 (2014): 4168–72.

Minor, Jesse, and Geoffrey A. Boyce. "Smokey Bear and the Pyropolitics of United States Forest Governance." *Political Geography* 62 (2018): 79–93.

Molyneux, David, and Dieudonné P. Sankara. "Guinea Worm Eradication: Progress and Challenges—Should We Beware of the Dog?" *PLoS Neglected Tropical Diseases* 11, no. 4 (2017): e0005495.

Mora, Camilo, Derek P. Tittensor, Sina Adl, Alastair G. B. Simpson, and Boris Worm. "How Many Species Are There on Earth and in the Ocean"?" *PLoS Biology* 9, no. 8 (2011): e1001127.

Mosca, Alexis, Marion Leclerc, and Jean P. Hugot. "Gut Microbiota Diversity and Human Diseases: Should We Reintroduce Key Predators in Our Ecosystem?" *Frontiers in Microbiology* 7 (2016): 455.

Moser, J. "Circuit Breakers." *Economic Perspectives* 14, no. 5 (September 1990).

Multilateral Trade Negotiations: The Uruguay Round. "Elaboration of US Agricultural Proposal with Respect to Food Security Submitted by the United States, Negotiating Group on Agriculture, Group of Negotiations on Goods (Gatt), Mtn.Gng/Ng5/W/61, 6 June 1988." 1988.

Munns, David P. D., and Kärin Nickelsen. "To Live Among the Stars: Artificial Environments in the Early Space Age." *History and Technology* 33, no. 3 (2017): 272–99.

National Research Council. *New Directions for Understanding Systemic Risk: A Report on a Conference Cosponsored by the Federal Reserve Bank of New York and the National Academy of Sciences*. Washington, D.C.: National Academies Press, 2007.

Nelson, Mark, and William F. Dempster. "Living in Space: Results from Biosphere 2's Initial Closure, an Early Testbed for Closed Ecological Systems on Mars." In *Strategies for Mars: A Guide to Human Exploration*, edited by American Astronautical Society, Carol R. Stoker, and Carter Emmart, 363–90. Science and Technology Series 86. San Diego, Calif.: Univelt, 1996.

Nelson, Mark, William F. Dempster, and John P. Allen. "Key Ecological Challenges for Closed Systems Facilities." *Advances in Space Research* 52, no. 1 (2013): 86–96.

Nelson, Mark, Nickolay S. Pechurkin, John P. Allen, Lydia A. Somova, and Josef I. Gitelson. "Closed Ecological Systems, Space Life Support and Biospherics." In *Environmental Biotechnology*, edited by Lawrence K. Wang, Volodymyr Ivanov, and Joo-Hwa Tay, 517–65. Handbook of Environmental Engineering 10. New York: Springer, 2010.

Netting, Robert McC. *Balancing on an Alp: Ecological Change and Continuity in a Swiss Mountain Community*. Cambridge: Cambridge University Press, 1981.

——. "What Alpine Peasants Have in Common: Observations on Communal Tenure in a Swiss Village." In *Case Studies in Human Ecology*, edited by Daniel G. Bates and Susan H. Lees, 219–31. New York: Springer, 1996.

Nettle, Daniel. "Explaining Global Patterns of Language Diversity." *Journal of Anthropological Archaeology* 17, no. 4 (1998): 354–74.

——. "Linguistic Diversity of the Americas Can Be Reconciled with a Recent Colonization." *Proceedings of the National Academy of Sciences* 96, no. 6 (1999): 3325–29.

Nettle, Daniel, and Suzanne Romaine. *Vanishing Voices: The Extinction of the World's Languages*. New York: Oxford University Press on Demand, 2000.

Nevala-Lee, A. "Asimov's Empire, Asimov's Wall." *Public Books*, January 7, 2020.

Newby, Gretchen, Adam Bennett, Erika Larson, Chris Cotter, Rima Shretta, Allison A. Phillips, and Richard G. A. Feachem. "The Path to Eradication: A Progress Report on the Malaria-Eliminating Countries." *Lancet* 387, no. 10029 (2016): 1775–84.

New York City Economic Development Council and Mayor's Office of Recovery and Resiliency. "Five Borough Food Flow: 2016 New York City Food Distribution and Resiliency Study Results." New York: New York

City Economic Development Council and Mayor's Office of Recovery and Resiliency, 2016.

Nicholas, Ralph W. "The Goddess Śītalā and Epidemic Smallpox in Bengal." *Journal of Asian Studies* 41, no. 1 (1981): 21–44.

Nichols, Theresa, Fikret Berkes, Dyanna Jolly, Norman B. Snow, and Community of Sachs Harbour. "Climate Change and Sea Ice: Local Observations from the Canadian Western Arctic." *Arctic* 57, no. 1 (2004): 68–79.

Nicolis, Gregoire, and " Ilya Prigogine. *Self-Organization in Non-Equilibrium Systems.* "New York: Wiley, 1977.

Nolan, Rachael H., Matthias M. Boer, Luke Collins, Víctor Resco de Dios, Hamish Clarke, Meaghan Jenkins, Belinda Kenny, and Ross A. Bradstock. "Causes and Consequences of Eastern Australia's 2019–20 Season of Mega-Fires." *Global Change Biology* 26, no. 3 (2020): 1039–1041.

Olmstead, Sheila, Carolyn Kousky, and Roger Sedjo. "Wildland Fire Suppression and Land Development in the Wildland/Urban Interface." Washington, D.C.: Resources for the Future, 2012.

Omi, Philip N. "Theory and Practice of Wildland Fuels Management." *Current Forestry Reports* 1, no. 2 (2015): 100–17.

Onisha, N. "Love of U.S. Food Shortening Okinawans' Lives/Life Expectancy Among Islands' Young Men Takes a Big Dive." *New York Times*, April 4, 2004.

Ostrom, Elinor. "Decentralization and Development: The New Panacea." In *Challenges to Democracy*, edited by Keith Dowding, James Hughes, and Helen Margetts, 237–56. London: Springer, 2001.

——. "Do Institutions for Collective Action Evolve"?" *Journal of Bioeconomics* 16, no. 1 (2014): 3–30.

——. *Governing the Commons: The Evolution of Institutions for Collective Action.* Cambridge: Cambridge University Press, 1990.

——. "A Long Polycentric Journey." *Annual Review of Political Science* 13 (2010): 1–23.

——. "Polycentric Systems for Coping with Collective Action and Global Environmental Change." *Global Environmental Change* 20, no. 4 (2010): 550–57.

Ostrom, Elinor, Thomas Dietz, Nives Dolšak, Paul C. Stern, Susan Stonich, and Elke U. Weber, eds. *The Drama of the Commons.* Washington, D.C.: National Academies Press, 2002.

Pagani, Giuliano Andrea, and Marco Aiello. "Power Grid Complex Network Evolutions for the Smart Grid." *Physica A: Statistical Mechanics and Its Applications* 396 (2014): 248–66.

Pandis, Spyros N., Ksakousti Skyllakou, Kalliopi Florou, Evangelia Kostenidou, Christos Kaltsonoudis, Erion Hasa, and Albert A. Presto. "Urban Particulate Matter Pollution: A Tale of Five Cities." *Faraday Discussions* 189 (2016): 277–90.

Parks, Sean A., Lisa M. Holsinger, Carol Miller, and Cara R. Nelson. "Wildland Fire as a Self-Regulating Mechanism: The Role of Previous Burns and Weather in Limiting Fire Progression." *Ecological Applications* 25, no. 6 (2015): 1478–92.

Pasteur, Louis. "Observations Relative a La Note De M. Duclaux." *Comptes rendus de l'Académie des Sciences* 100 (1885): 68.

Pastor-Satorras, Romualdo, Claudio Castellano, Piet Van Mieghem, and Alessandro Vespignani. "Epidemic Processes in Complex Networks." *Reviews of Modern Physics* 87, no. 3 (2015): 925.

Patel, Raj, and Philip McMichael. "A Political Economy of the Food Riot." *Review: A Journal of the Fernand Braudel Center* 32, no. 1 (2009): 9–35.

Patrouch, Joseph F. *The Science Fiction of Isaac Asimov*. New York: Doubleday, 1974.

Pelkey, James. Entrepreneurial Capitalism and Innovation: A History of Computer Communications 1968–1988." 2007. www. historyofcomputer communications.

Pellegrini, Pablo A., and Galo E. Balatti. "Noah's Arks in the XXI Century. A Typology of Seed Banks." *Biodiversity and Conservation* 25, no. 13 (2016): 2753–69.

Penczykowski, Rachel M., Anna-Liisa Laine, and Britt Koskella. "Understanding the Ecology and Evolution of Host–Parasite Interactions Across Scales." *Evolutionary Applications* 9, no. 1 (2016): 37–52.

Penn, I., P. Eavis, and J. Glanz. "How PG&E Ignored Fire Risks in Favor of Profits." *New York Times*, March 19, 2019.

Pimm, Stuart L., Clinton N. Jenkins, Robin Abell, Thomas M. Brooks, John L. Gittleman, Lucas N. Joppa, Peter H. Raven, et al. "The Biodiversity of Species and Their Rates of Extinction, Distribution, and Protection." *Science* 344, no. 6187 (2014): 1246752.

Pingali, P. "The Green Revolution and Crop Diversity." In *Routledge Handbook of Agricultural Biodiversity*, edited by Danny Hunter, Luigi Guarino, Charles Spillane, and Peter C. McKeown, chap. 12. Abingdon, U.K.: Routledge, 2017.

Pingali, Prabhu L. "Green Revolution: Impacts, Limits, and the Path Ahead." *Proceedings of the National Academy of Sciences* 109, no. 31 (2012): 12302–308.

Plotkin, Stanley A. "Vaccines: Past, Present and Future." *Nature Medicine* 11, no. S4 (2005): S5.

Pomeroy, Robert S., and Fikret Berkes. "Two to Tango: The Role of Government in Fisheries Co-Management." *Marine Policy* 21, no. 5 (1997): 465–80.

Popkin, Barry M. "Global Nutrition Dynamics: The World Is Shifting Rapidly Toward a Diet Linked with Noncommunicable Diseases." *American Journal of Clinical Nutrition* 84, no. 2 (2006): 289–98.

Porter, Jane M., and Douglas E. Bowers. "A Short History of US Agricultural Trade Negotiations." Washington, D.C.: U.S. Department of Agriculture–Economic Research Service, 1989.

Portugal, Steven J., Tatjana Y. Hubel, Johannes Fritz, Stefanie Heese, Daniela Trobe, Bernhard Voelkl, Stephen Hailes, Alan M. Wilson, and James R. Usherwood. "Upwash Exploitation and Downwash Avoidance by Flap Phasing in Ibis Formation Flight." *Nature* 505, no. 7483 (2014): 399.

Price, Charles A., and Joshua S. Weitz. "Costs and Benefits of Reticulate Leaf Venation." *BMC Plant Biology* 14, no. 1 (2014): 234.

Puma, Michael J., Satyajit Bose, So Young Chon, and Benjamin I. Cook. "Assessing the Evolving Fragility of the Global Food System." *Environmental Research Letters* 10, no. 2 (2015): 024007.

Pyne, Stephen J. "Firestick History." *Journal of American History* 76, no. 4 (1990): 1132–41.

Radetsky, Michael. "Smallpox: A History of Its Rise and Fall." *Pediatric Infectious Disease Journal* 18, no. 2 (1999): 85–93.

Rajasekharan, P. E. "Gene Banking for ex Situ Conservation of Plant Genetic Resources." In *Plant Biology and Biotechnology*, Vol. 2, *Plant Genomics and Biotechnology*, edited by Bir Bahadur, Manchikatla Venkat Rajam, Leela Sahijram, and K. V. Krishnamurthy, 445–59. New Delhi: Springer, 2015.

Rajoka, Muhammad Shahid Riaz, Junling Shi, Hafiza Mahreen Mehwish, Jing Zhu, Qi Li, Dongyan Shao, Qingsheng Huang, and Hui Yang. "Interaction Between Diet Composition and Gut Microbiota and Its Impact on Gastrointestinal Tract Health." *Food Science and Human Wellness* 6, no. 3 (2017): 121–30.

Rascovar, B. "Shipping Pioneer Largely Ignored." *The Sun*, June 14, 2001.

Rayner, Anthony J., K. A. Ingersent, and R. C. Hine. "Agriculture in the Uruguay Round: An Assessment." *Economic Journal* 103, no. 421 (1993): 1513–27.

Resnick, B., U. Irfan, and S. Samual. "8 Things Everyone Should Know About Australia's Wildlife Disaster." *Vox*, January 22, 2020.

Reyes-Velarde, A. "California's Camp Fire Was the Costliest Global Disaster Last Year, Insurance Report Shows." *Los Angeles Times*, January 11, 2019.

Reynolds, Craig W. "Flocks, Herds and Schools: A Distributed Behavioral Model." In *SIGGRAPH '87: Proceedings of the 14th Annual Conference on Computer Graphics and Interactive Techniques*, edited by Maureen C. Stone, 25–34. New York: Association for Computing Machinery, 1987.

Ribot, Jesse C., Arun Agrawal, and Anne M. Larson. "Recentralizing While Decentralizing: How National Governments Reappropriate Forest Resources." *World Development* 34, no. 11 (2006): 1864–86.

Ritchie, Mark. *Impact of Gatt on World Hunger*. Minneapolis: Institute for Agriculture and Trade Policy, 1988.

Ronfeldt, David. "In Search of How Societies Work." RAND Pardee, 2007.

Rosengaus, Rebeca B., James F. A. Traniello, Tammy Chen, Julie J. Brown, and Richard D. Karp. "Immunity in a Social Insect." *Naturwissenschaften* 86, no. 12 (1999): 588–91.

Rothman, Hal K. *A Test of Adversity and Strength: Wildland Fire in the National Park System*. Washington, D.C.: National Park Service, 2005.

Roth-Nebelsick, Anita, Dieter Uhl, Volker Mosbrugger, and Hans Kerp. "Evolution and Function of Leaf Venation Architecture: A Review." *Annals of Botany* 87, no. 5 (2001): 553–66.

Saavedra, Serguei, Felix Reed-Tsochas, and Brian Uzzi. "A Simple Model of Bipartite Cooperation for Ecological and Organizational Networks." *Nature* 457, no. 7228 (2009): 463–66.

Sack, Lawren, Elisabeth M. Dietrich, Christopher M. Streeter, David Sánchez-Gómez, and N. Michele Holbrook. "Leaf Palmate Venation and Vascular Redundancy Confer Tolerance of Hydraulic Disruption." *Proceedings of the National Academy of Sciences* 105, no. 5 (2008): 1567–72.

Sack, Lawren, and Christine Scoffoni. "Leaf Venation: Structure, Function, Development, Evolution, Ecology and Applications in the Past, Present and Future." *New Phytologist* 198, no. 4 (2013): 983–1000.

Sagarin, Rafe. *Learning from the Octopus: How Secrets from Nature Can Help Us Fight Terrorist Attacks, Natural Disasters, and Disease*. New York: Basic Books, 2012.

Schapiro, M. "How Seeds from War-Torn Syria Could Help Save American Wheat." *YaleEnvironment360*, May 14, 2018.

Schiffman, R. "The Seeds of the Future." *New Scientist* 225, no. 3002 (2015): 23.

Schindler, Daniel E., Jonathan B. Armstrong, and Thomas E. Reed. "The Portfolio Concept in Ecology and Evolution." *Frontiers in Ecology and the Environment* 13, no. 5 (2015): 257–63.

Seabrook, J. "Sowing for Apocalypse: The Quest for a Global Seed Bank." *New Yorker*, August 20, 2007.

Seebacher, Frank. "The Evolution of Metabolic Regulation in Animals." *Comparative Biochemistry and Physiology Part B: Biochemistry and Molecular Biology* 224 (2017): 195–203.

Seto, Karen C., Steven J. Davis, Ronald B. Mitchell, Eleanor C. Stokes, Gregory Unruh, and Diana Ürge-Vorsatz. "Carbon Lock-In: Types, Causes, and Policy Implications." *Annual Review of Environment and Resources* 41 (2016): 425–52.

Severinghaus, Jeffrey P., Wallace S. Broecker, William F. Dempster, Taber MacCallum, and Martin Wahlen. "Oxygen Loss in Biosphere 2." *EOS, Transactions, American Geophysical Union* 75, no. 3 (1994): 33–37.

Shchelkunov, Sergei N. "Emergence and Reemergence of Smallpox: The Need for Development of a New Generation Smallpox Vaccine." *Vaccine* 29 (2011): D49–D53.

Shen, Zhuang, Fang Ning, Weigong Zhou, Xiong He, Changying Lin, Daniel P. Chin, Zonghan Zhu, and Anne Schuchat. "Superspreading SARS Events, Beijing, 2003." *Emerging Infectious Diseases* 10, no. 2 (2004): 256.

Simmons, Brian J., Leyre A. Falto-Aizpurua, Robert D. Griffith, and Keyvan Nouri. "Smallpox: 12 000 Years from Plagues to Eradication: A Dermatologic Ailment Shaping the Face of Society." *JAMA Dermatology* 151, no. 5 (2015): 521.

Simone-Finstrom, Michael. "Social Immunity and the Superorganism: Behavioral Defenses Protecting Honey Bee Colonies from Pathogens and Parasites." *Bee World* 94, no. 1 (2017): 21–29.

Singel, R. "Vint Cerf: We Knew What We Were Unleashing on the World." *Wired*, April 23, 2012.

Sloand, Elaine M., Elisabeth Pitt, and Harvey G. Klein. "Safety of the Blood Supply." *JAMA* 274, no. 17 (1995): 1368–73.

Smil, Vaclav. *Cycles of Life: Civilization and the Biosphere*. Scientific American Library. New York: Freeman, 1996.

Smith, Adam. *The Wealth of Nations: An Inquiry Into the Nature and Causes of the Wealth of Nations*. Petersfield, U.K.: Harriman House, 2010.

Smith, Jordan Fisher. "Life Under the Bubble." *Discover*, December 19, 2010.

Sommer, Morten O. A., and Gautam Dantas. "Antibiotics and the Resistant Microbiome." *Current Opinion in Microbiology* 14, no. 5 (2011): 556–63.

Sonestedt, Emily, Nina Cecilie Øverby, David E. Laaksonen, and Bryndis Eva Birgisdottir. "Does High Sugar Consumption Exacerbate Cardiometabolic Risk Factors and Increase the Risk of Type 2 Diabetes and Cardiovascular Disease?" *Food & Nutrition Research* 56, no. 1 (2012): 19104.

Sousa Araujo, T. de, J. de Melo, W. Junior, and U. Albuquerque. "Medicinal Plants." In *Introduction to Ethnobiology*, edited by U. Albuquerque and R. Alves, 143–49. Cham, Switzerland: Springer, 2016.

Stark, K. "Climate Change Is Driving California Wildfires. The Kincaid Fire? No So Much." *KQED Science*, November 6, 2019.

Stephens, P., and H. Leufkens. "The World Medicines Siutation 2011: Research and Development"," edited by World Health Organization, 1–28. Geneva: World Health Organization, 2011.

Stroeymeyt, Nathalie, Barbara Casillas-Pérez, and Sylvia Cremer. "Organisational Immunity in Social Insects." *Current Opinion in Insect Science* 5 (2014): 1–15.

Strogatz, Steven H. "Exploring Complex Networks." *Nature* 410, no. 6825 (2001): 268–76.

Strullu-Derrien, C., M.-A. Selosse, P. Kenrick, and F. Martin. "The Origin and Evolution of Mycorrhizal Symbiosis: From Palaeomycology to Phylogenomics." *New Phytologist* 220, no. 4 (2018): 1012–30.

Subrahmanyam, Avanidhar. "Algorithmic Trading, the Flash Crash, and Coordinated Circuit Breakers." *Borsa Istanbul Review* 13, no. 3 (2013): 4–9.

Sumpter, David J. T. *Collective Animal Behavior*. Princeton, N.J.: Princeton University Press, 2010.

Tainter, Joseph. *The Collapse of Complex Societies*. Cambridge: Cambridge University Press, 1988.

Tainter, Joseph A. "Social Complexity and Sustainability." *Ecological Complexity* 3, no. 2 (2006): 91–103.

Tan, Siang Yong, and Yvonne Tatsumura. "Alexander Fleming (1881–1955): Discoverer of Penicillin." *Singapore Medical Journal* 56, no. 7 (2015): 366.

Taylor, S. Ross, and Scott M. McLennan. "The Evolution of Continental Crust." *Scientific American* 274, no. 1 (1996): 76–81.

Thorburn, A. Lennox. "Paul Ehrlich: Pioneer of Chemotherapy and Cure by Arsenic (1854–1915)." *Sexually Transmitted Infections* 59, no. 6 (1983): 404–405.

Tilly, Louise A. "The Food Riot as a Form of Political Conflict in France." *Journal of Interdisciplinary History* 2, no. 1 (1971): 23–57.

Toole, Michael J. "So Close: Remaining Challenges to Eradicating Polio." *BMC Medicine* 14, no. 1 (2016): 43.

Tootell, Betty." *All Four Engines Have Failed; the True and Triumphant Story of Flight BA 009 and the Jakarta Incident*." Auckland: Hutchinson Group, 1985.

Torero, Maximo. "Alternative Mechanisms to Reduce Food Price Volatility and Price Spikes: Policy Responses at the Global Level." In *Food Price Volatility and Its Implications for Food Security and Policy*, edited by Matthias Kalkuhl, Joachim von Braun, and Maximo Torero, 115–38. Cham, Switzerland: Springer, 2016.

Traniello, James F. A., Rebeca B. Rosengaus, and Keely Savoie. "The Development of Immunity in a Social Insect: Evidence for the Group Facilitation of Disease Resistance." *Proceedings of the National Academy of Sciences* 99, no. 10 (2002): 6838–42.

Travers, Jeffrey, and Stanley Milgram. "The Small World Problem." *Psychology Today* 1 (1967): 61–67.

Turner, B. L. "The Ancient Maya: Sustainability and Collapse?" In *Routledge Handbook of the History of Sustainability*, edited by Jeremy L. Caradonna, 57–68. Abingdon, U.K.: Routledge, 2017.

Tyers, Rod. "The Cairns Group and the Uruguay Round of International Trade Negotiations." *Australian Economic Review* 26, no. 1 (1993): 49–60.

Umpleby, Stuart. "A Short History of Cybernetics in the United States." *Österreichische Zeitschrift für Geschichtswissenschaften* 19, no. 4 (2008): 28.

United Nations. "Paris Agreement, Chapter XXVII, 7.D Paris Agreement, 12 December 2015." 2015.

——. "United Nations Framework Convention on Climate Change." FCCC/INFORMAL/84 GE.05-62220 (E) 200705, 1992.

—— "We the Peoples: The Role of the United Nations in the Twenty-First Century. Report of the Secretary-General, a/54/2000, 27 March 2000." New York: United Nations General Assembly, 2000.

Unruh, Gregory C. "Understanding Carbon Lock-In." *Energy Policy* 28, no. 12 (2000): 817–30.

U.S. Securities Exchange Commission. "Findings Regarding the Market Events of May 6, 2010: Report to the Staffs of the Cftc and Sec to the Joint Advisory Committee on Emerging Regulatory Issues." New York: U.S. Securities Exchange Commission, 2010.

Vallée, Robert. "History of Cybernetics." In *Systems Science and Cybernetics*, vol. 3, ed. F. Parra-Luna (Oxford: Eolss, 2009): 22–33.

Van Valen, Leigh. "A New Evolutionary Law." *Evolutionary Theory* 1 (1973): 1–30.

Vavilov, Nikolaĭ Ivanovich, Mykola I. Vavylov, Níkolaj Ívanovítsj Vavílov, and Vladimir Filimonovich Dorofeev. *Origin and Geography of Cultivated Plants*. Cambridge: Cambridge University Press, 1992.

Von Neumann, John. "Probabilistic Logics and the Synthesis of Reliable Organisms from Unreliable Components." *Automata Studies* 34 (1956): 43–98.

Wade, L. "How Syrians Saved an Ancient Seedbank from Civil War." *Wired*, April 17, 2015.

Walker, Brian, Crawford S. Holling, Stephen R. Carpenter, and Ann Kinzig. "Resilience, Adaptability and Transformability in Social–Ecological Systems." *Ecology and Society* 9, no. 2 (2004): 5.

Walker, James C. G., P. B. Hays, and James F. Kasting. "A Negative Feedback Mechanism for the Long-Term Stabilization of Earth's Surface Temperature." *Journal of Geophysical Research: Oceans* 86, no. C10 (1981): 9776–82.

Wall, H. "Stephen Hawking Never Reached Space, but He Sought to Lift All of Humanity." *Space.com*, March 14, 2018.

Wang, Li, Fengying Zhang, Eva Pilot, Jie Yu, Chengjing Nie, Jennifer Holdaway, Linsheng Yang, et al. "Taking Action on Air Pollution Control in the Beijing–Tianjin–Hebei (Bth) Region: Progress, Challenges and Opportunities." *International Journal of Environmental Research and Public Health* 15, no. 2 (2018): 306.

Wang, Zhen, Yamir Moreno, Stefano Boccaletti, and Matjaž Perc. "Vaccination and Epidemics in Networked Populations—an Introduction." *Chaos, Solitons, and Fractals* 103 (2017): 177–83.

Ward-Perkins, Bryan. *The Fall of Rome: And the End of Civilization*. Oxford: Oxford University Press, 2006.

Watts, Duncan. "A Simple Model of Global Cascades on Random Networks." *Proceedings of the National Academy of Sciences* 99, no. 9 (2002): 5766-71.

Watts, Duncan. *Six Degrees: The Science of a Connected Age*. New York: Norton, 2003.

Watts, Duncan J., and Steven H. Strogatz. "Collective Dynamics of 'Small-World' Networks." *Nature* 393, no. 6684 (1998): 440–42.

Weiss, Madeline C., Filipa L. Sousa, Natalia Mrnjavac, Sinje Neukirchen, Mayo Roettger, Shijulal Nelson-Sathi, and William F. Martin. "The Physiology and Habitat of the Last Universal Common Ancestor." *Nature Microbiology* 1, no. 9 (2016): 16116.

Wellesley, Laura, Felix Preston, Johanna Lehne, and Rob Bailey. "Chokepoints in Global Food Trade: Assessing the Risk." *Research in Transportation Business & Management* 25 (2017): 15–28.

Werrell, Caitlin E., Francesco Femia, and Troy Sternberg. "Did We See It Coming? State Fragility, Climate Vulnerability, and the Uprisings in Syria and Egypt." *SAIS Review of International Affairs* 35, no. 1 (2015): 29–46.

"When Volcanic Ash Stopped a Jumbo at 37.000ft." *BBC News Magazine*, April 15, 2010.

White, Michael. *Isaac Asimov: A Life of the Grand Master of Science Fiction.* New York: Carroll & Graf, 2005.

Wiener, Norbert. *Cybernetics or Control and Communication in the Animal and the Machine.* 2nd ed. Cambridge, Mass.: MIT Press, 1965.

Wiggins, Steve, and Sharada Keats. "Looking Back, Peering Forward: Food Prices and the Food Price Spike of 2007/08." London: Overseas Development Institute, March 28, 2013. www. odi. org/sites/odi. org. uk/files/odi-assets /publications-opinion-files/8339. pdf.

Wilkins, E. T. "Air Pollution and the London Fog of December, 1952." *Journal of the Royal Sanitary Institute* 74, no. 1 (1954): 1–21.

Willcox, D. Craig, Bradley J. Willcox, Hidemi Todoriki, and Makoto Suzuki. "The Okinawan Diet: Health Implications of a Low-Calorie, Nutrient-Dense, Antioxidant-Rich Dietary Pattern Low in Glycemic Load." *Journal of the American College of Nutrition* 28, no. S4 (2009): 500S–516S.

Willcox, Donald Craig, Giovanni Scapagnini, and Bradley J. Willcox. "Healthy Aging Diets Other Than the Mediterranean: A Focus on the Okinawan Diet." *Mechanisms of Ageing and Development* 136 (2014): 148–62.

Williams, Alan N., Scott D. Mooney, Scott A. Sisson, and Jennifer Marlon. "Exploring the Relationship Between Aboriginal Population Indices and Fire in Australia over the Last 20,000 Years." *Palaeogeography, Palaeoclimatology, Palaeoecology* 432 (2015): 49–57.

Wolfe, Nathan D., Claire Panosian Dunavan, and Jared Diamond. "Origins of Major Human Infectious Diseases." *Nature* 447, no. 7142 (2007): 279.

Wood, Stephen A., Matthew R. Smith, Jessica Fanzo, Roseline Remans, and Ruth S. DeFries. "Trade and the Equitability of Global Food Nutrient Distribution." *Nature Sustainability* 1, no. 1 (2018): 34.

World Health Assembly. "Thirty-Third World Health Assembly, Geneva 5–23 May 1980: Resolutions and Decisions, Annexes." Geneva: World Health Organization, 1980.

——. "Eighth World Health Assembly: Mexico D.F., 10–27 May 1955." Geneva: World Health Organization, 1955.

World Health Organization. "Operational Framework for the Deployment of the World Health Organization Smallpox Vaccine Emergency Stockpile in Response to a Smallpox Event." Geneva: World Health Organization, 2017.

——. "SARS: Lessons from a New Disease." In *The World Health Report 2003—Shaping the Future*, 73–82. Geneva: World Health Organization, 2005.

Worobey, Michael, Thomas D. Watts, Richard A. McKay, Marc A. Suchard, Timothy Granade, Dirk E. Teuwen, Beryl A. Koblin, et al. "1970s and 'Patient 0' HIV-1 Genomes Illuminate Early HIV/AIDS History in North America." *Nature* 539, no. 7627 (2016): 98–101.

Wright, P. "Colorado Blaze Continues to Grow; Entire San Juan National Forest to Close." *The Weather Channel*, 2018.

Wright, R. T., and B. J. Nebel. *Environmental Science: Toward a Sustainable Future.* 8th ed. Upper Saddle River, N.J.: Prentice Hall, 2000.

Yatsunenko, Tanya, Federico E. Rey, Mark J. Manary, Indi Trehan, Maria Gloria Dominguez-Bello, Monica Contreras, Magda Magris, et al. "Human Gut Microbiome Viewed Across Age and Geography." *Nature* 486, no. 7402 (2012): 222.

Yerxa, Donald A. "An Interview with Bryan Ward-Perkins on the Fall of Rome." *Historically Speaking* 7, no. 4 (2006): 31–33.

Yong, E. "After Last Year's Hurricanes, Caribbean Lizards Are Better at Holding on for Dear Life." *Atlantic*, July 25, 2018.

INDEX